ELECTRIC LIVING

The Powerful Life!

By

Kolie E. Crutcher III

KOLIE CRUTCHER PUBLICATIONS, INC.
Power Reading for Power People™

Kolie Crutcher Enterprises, Inc.
New York, NY
www.ElectricLiving.net

ISBN 978-0-9814643-0-5

Published by Kolie Edward Crutcher III, New York; 2008, 2009

Printed in the United States of America on acid-free paper.

CONTENTS

INTRODUCTION vii

A PROMISE FROM THE AUTHOR ix

FOREWORD xi

AND THE OSCAR GOES TO.................................... xv

HOW TO READ THIS BOOK xxiii

PART I

CHAPTER 1 LIGHT – Your Realized Dream 3

CHAPTER 2 ELECTRIC CURRENT – You *in Action!* 13

CHAPTER 3 VOLTAGE SOURCE – Your Source of
Motivation 23

CHAPTER 4 RESISTANCE – Fear 33

CHAPTER 5 ELECTRIC CIRCUIT – The Planned Path
to Your Realized Dream 47

CHAPTER 6 POWER – Action Despite Fear 57

PART II

CHAPTER 7 INSULATION – Self-Discipline 69

CHAPTER 8 CONDUCTIVITY - Faith 79

CHAPTER 9 SHORT CIRCUIT - Complacency/
Conformity/ Comfort Zone 89

CHAPTER 10 OPEN CIRCUIT - Paralyzed by Fear 97

CHAPTER 11 CAPACITANCE - Imagination 107

CHAPTER 12 INDUCTANCE - Persuasion 117

CHAPTER 13 FREQUENCY – Attitude! 127

CONCLUSION – COMPLETING THE
CIRCUIT OF LIFE 145

APPENDIX I 149
APPENDIX II..157
APPENDIX III 159

INTRODUCTION

By Monroe Mann
Founder, Unstoppable Artists, LLC
CEO, Loco Dawn Films, LLC
Combat Veteran, Operation Iraqi Freedom
Author, *The Theatrical Juggernaut, Battle Cries for the Underdog,* etc.

It is very rare to find a book that not only inspires, but also inspires one to think about a subject in an entirely novel and original way. Kolie Crutcher has done exactly this with his groundbreaking book, *Electric Living: The Powerful Life!*

Kolie is an amazing person, and for a variety of reasons. The first thing you notice about this man is his bright glowing smile. Once you hang out with him for a while, you start to realize that his smile isn't the only part of him that glows: his heart, passion, and zest for life are unquenchable. And once you read this book, you'll realize that his insight and creativity are second to none. Truly, anyone with the ability to draw together two seemingly unrelated subjects—motivation and electricity—and develop a book that leaves you with a better understanding of both, simply must be commended.

As a motivational author and speaker myself, as well as an FCC licensed amateur (Ham) radio operator familiar with electrical terms, principles, and circuits, I can unequivocally say that you are going to love this book. If you're looking for assistance on your road to greatness, or even looking for a novel way to learn the basics of electricity, this book is the ticket. Moving gracefully from electrical principle to electrical principle—and explaining clearly how each principle can be applied to the practical realization of your dreams—is no small feat. Mr. Crutcher has done just that, and in a manner that even the electrical layman can understand.

If you are reading this introduction in a bookstore… don't hesitate, and purchase this book today. You'll not only enjoy reading it the first time, but you'll also enjoy reading it a second, third, and fourth time. *Electric Living* is one of those rare books that gets better and better each time you read it—not only do you better understand the principles as you re-read the book, but we all need an 'electrical recharge' on a regular basis… and this book will give you those indispensable jumpstarts.

I can't say enough about this book. The entire concept is genius. The presentation is exhilarating. The explanation is clear and concise. Truly, the fact that Kolie was able to codify this 'lofty notion' into a book that is easy for everyone to understand is beyond applaud. To say that I am impressed with this book is an understatement.

Electric Living is sure to become a motivational classic. It already is in my mind.

A PROMISE FROM THE AUTHOR

Probably the greatest tragedy of the human race is the failure of the overwhelming majority of people to realize their dreams.

Your dreams are realized as a result of creating power in your life. But, because most people have no clue about how to create power in their lives, their dreams remain unrealized. With that being said, quite naturally the question becomes, "How do I create power in my life?"

The answer to this question is found within the pages of this book. Think about the following:

> **You cannot turn on the light in your home without electric power.**

> **In the same fashion, you cannot turn on the dreams in your life without personal power.**

Electric Living: *The Powerful Life!* demonstrates the process of creating the personal power necessary to turn on the dreams in your life by relating it to the electric power necessary to turn on a common light bulb in your home.

That is my promise to you, the reader. **This is NOT just another motivational book!** And here is why: as you will learn, motivation is only *part* of the formula for creating power. If all you want is motivation, there are countless other books that you can purchase. Yes, **Electric Living: *The Powerful Life!*** *will* motivate you, but it does not stop there. This book goes the extra mile and gives you a formula for creating power—that personal power necessary to realize the dreams in your life!

Kolie E. Crutcher III

FOREWORD

If a man is called to be a street sweeper, he should sweep streets even as Michelangelo painted, or Beethoven composed music, or Shakespeare wrote poetry. He should sweep streets so well that all the host of heaven and earth will pause to say, here lived a great street sweeper who did his job well.

– Martin Luther King, Jr.

Have you ever been in a storm, and the lights went out in your house? Whoa! Not a good feeling, especially at night.

What became the #1 priority? You guessed it, turning on the lights. While the light was out, you were lost, you bumped into things, and you were easily fooled. Now think about the exact moment when those lights came back on. What was the first thought in your mind? "Yeah! SUCCESS!"

But let's hold on to that thought for one second. Can we dig a little deeper? Can wc, by understanding some basic principles of electricity, understand *how* we achieve the small-scale success of turning on a light bulb? And can we, by understanding the small-scale success of turning on a light bulb, begin to understand how to achieve the large-scale success in our lives of turning on our dreams?

The answer is yes. Up above, when you thought, "Yeah! SUCCESS!", you were absolutely right. Turning on the light bulb IS SUCCESS!

You see, we need light! Without light of some form or another, it is impossible for us to see. The human eye converts light into electrical impulses that the brain interprets as vision. Have you ever thought about the process of turning on a light bulb? Have you ever thought about how a person becomes successful in life? **HAVE YOU EVER THOUGHT THAT THE SAME PRINCIPLES THAT LEAD TO THE "SUCCESS" OF TURNING ON A LIGHT BULB ARE THE SAME PRINCIPLES THAT LEAD TO THE "SUCCESS" OF REALIZING THE DREAMS IN YOUR LIFE?**

TURNING ON THE LIGHT *IS* SUCCESS IN LIFE!!!

Earl Nightingale once described success as "the progressive realization of a worthy ideal." Stated another way, the successful person is a person who is moving toward a predetermined goal or dream because that's what he or she *chose* to do—deliberately. A success is the man who owns a corner deli because that is his dream—that's what he *wanted* to do. A success is the schoolteacher who teaches students because that's what she *wanted* to do. A success is the amputee who was told he would never walk again, but finished the 26.2 miles of the New York City Marathon because he *chose* to do so. A success is the cash poor young man who turns himself into a billionaire businessman because that is what he *wanted* to do.

Success for one person, you realize, may not be success for another person. If a person becomes a surgeon, he may be viewed as a success by those around him. They may think, "Wow, think of all the hard work, dedication, and studying needed to become a surgeon!"

This may very well be true. However, success cannot be defined by others. Success can only be defined by you. The surgeon may have only put in the hard work, dedication, and studying to be a surgeon only because he felt like that was what he was "supposed" to do. Deep down, this surgeon may have actually desired to be a cartoon artist. But maybe someone told him that being a cartoon artist is not a "respectable" enough career, or that it was not an "important" field. By Nightingale's definition, this surgeon in not a success because he is not in the line of work he truly desires.

This is not to say he is a bad surgeon. He may be an excellent surgeon, but he will not reach his full potential because his heart is not 100% dedicated to his line of work. Being a surgeon was not his dream. His dream as a kid was to create a brand new comic book and draw all of the characters himself. He would draw and doodle for hours on end as a kid, using his imagination to make his characters life-like, even though they existed only on paper and in his mind. He would envision his cartoons as super heroes who were starring in their own Saturday morning cartoon. He would see his action figures in toy stores across the nation. He even envisioned floats depicting his characters in the Macy's Thanksgiving Day Parade. That was his dream. A dream he chose not to follow.

You see, true success is something that comes from within. When you have a strong desire to do or become something, there is a reason for that desire. It is not by accident. The fact that you may be young or inexperienced does not matter because you are guided by your spirit which always leads you down the right path, *if you listen to it.* The problems start coming in as we get older and start to become influenced by others, who are so called "wiser".

While it may be true that other people can tell you much information about a particular subject, they have no expertise about your special path in life. No one can truly know how to walk that path except you. You must make that decision based on what you want out of life, not what others want for you.

Electric Living: *The Powerful Life!* is a book designed to help you take the principles of electricity, apply them to a successful life, and thus help you find your *own* way through the maze of life's circuit board. And if you apply what I offer, I assure you that far from your life short circuiting; it will glow more brightly than an entire city's power system.

AND THE OSCAR GOES TO...

Yes, ladies and gentlemen, this is it! A small sample of all the people I would like to thank who helped to make this all possible.

Most of these acknowledgements are in no particular order, but of course I'd like to thank God first (for obvious reasons), and myself second (this book didn't write itself).

Special thanks goes out to Monroe Mann, who was with me step by step, offering swift kicks in the butt when they were needed; wise counsel, advice, and editing, when warranted; and acting as my business coach and advisor: Monroe, I really appreciate it all! You, your advisors, your business school *Unstoppable Artists (www. UnstoppableArtists.com),* and your first two books, *The Theatrical Juggernaut – The Psyche of the Star,* and *Battle Cries for the Underdog – Fightin' Words for an Extraordinary Life,* have inspired me immensely along the way. They have provided a great deal of focus throughout this whole process and I wouldn't be here without your help, guidance, and expertise. Your fire burns white hot, you are a true inspirational force, and echoes of you reverberate through this book, as anyone who knows you or has read your books will notice (i.e., the acknowledgements for one!). Thank you for never considering my feelings, but instead kicking the excuses out of me and showing me how to kick them out of myself. You are the true embodiment of someone living electrically; the power within your soul is enough to run an entire city; maybe even an entire country.

I'd like to thank my Mom, Essie, and my Dad, Colie Jr. for being there for me. I always know I can depend on you guys whenever I need anything.

I also want to thank my dad for telling me to "watch out for those 'shysters'" and telling me about the story of the goat that fell in the ditch*.

* See p. xix

Mom, thank you for trying to instill in me the need to always do the right thing. I know many times it seems like I do the exact opposite of what you have taught me, but it really is for the right purpose. I have never forgotten what you told me. Thank you Grandma Wense for being the best cook in the world, and for your resilient spirit.

I'd like to thank my two sisters, Dai Dai and Tamara for making my role as "Big Brother" so rewarding. Thank you Dai Dai for all your "counseling" and analysis of my "aggressive tendencies". Thank you Tam for all those colorful pictures you sent me and cartoons you drew. They always made me smile. Tamara will probably want a whole page for herself in my next book. Maybe it would be labeled "Luves, Misses, Kisses!"

I'd like to thank my deceased grandparents (Colie Sr., Charlie, and Pandora) for helping mold my work ethic and desire to stand on my own. Although you are missed, you are not forgotten—your spirit remains.

I'd like to thank all my fraternity "bruhs" of Kappa Alpha Psi Fraternity, Incorporated (especially Eta Upsilon and Miami Alumni Chapters) for engraining our fraternity's fundamental purpose of "ACHIEVEMENT in every field of human endeavor."

I'd like to thank Ayana (M.O.A.O.A) for all your loving encouragement and for always being by my side. Your support in all my endeavors is truly appreciated, and you are always on my mind and in my heart!

I'd like to thank the American Academy of Dramatic Arts in New York City. The training I received and the relationships I formed at 120 Madison Avenue in NYC will never be forgotten. A "big up" goes out to the greatest professional acting school in the world.

I'd like to thank the faculty and staff of professors at Mississippi State University for introducing me to the electrical principles that this book is founded upon.

I'd like to thank my pastor Apostle Alton R. Williams of World Overcomers Outreach Ministries Church in Memphis, TN. You baptized me when I was 8 years old, and your strong will and words of wisdom touch me in a way different than any other pastor.

I'd like to thank Jorge and Jinet for giving me a place to stay when I first came to the Bronx in the dead of winter with no money, no job, and no references. You were the only ones who gave me an opportunity to have a roof over my head without having a job, deposits, and references. If it were not for you guys trusting in me, I would have been out in the streets.

I'd like to thank "The Basement" in Boston. At the time, those dark ice cold nights down there seemed like they were taking so much out of me. I didn't know it at the time, but they were actually putting something *into* me. I gained so much in the way of the art of survival. I have never been so cold in my life, and when you're broke, it's even colder.

I'd like to thank me again (remember, I did write this book).

I'd like to thank Napoleon Hill, whose inspirational classic *Think and Grow Rich* set the stage for my understanding of "The Secret". Many of the great inspirational books of today have their foundation in the principles outlined by Hill in *Think and Grow Rich*.

I'd like to thank Terrence Howard, because although we've never met, you have been such an inspiration to me, and I look forward to working with you in the near future. You are hands down my favorite actor (besides myself), and I hope by the time this book comes out, we will be starring together in *Hustle & Flow 2*. I know this movie is coming; it's just a matter of time.

I'd like to thank everyone who ever wanted me to fail. Your "hating" has lifted me to new heights and has rekindled my fire to succeed. So to you "haters", a special message is warranted: "DON'T HATE, CONGRATULATE!" Along those same lines, I would like to thank in advance those of you who pretend to be on my side, but really wanted me to fail. You "shysters" (as my dad would call them) are just flat out of luck.

I'd like to thank Mr. Vincent Dial for teaching me that "Practice does not make perfect. PERFECT practice makes perfect." And I'd like to thank Mr. Gregory Simmons for teaching me that "Practice makes Permanent."

I'd like to thank Three 6 Mafia, Eightball & MJG, Dipset, Jodeci, Yo Gotti, Alicia Keys, Bon Jovi, Jay-Z, Tupac, Notorious B.I.G., Sean "Diddy" Combs, Master P, Damon Dash, Jadakiss, Mary J. Blige, 50-Cent, T.I., Young Jeezy, Lil' Wayne, and all of the various recording artists I listen to. Even though I have never met any of you guys in person, your music has inspired me in ways you could not imagine.

I'd like to thank the saleslady at the Chevrolet dealership in Memphis, who, when I wanted to test drive a Corvette, simply gave me the keys and said, "Enjoy!" I had a 3-hour "test-drive", and confirmed my thoughts of the necessity of having one.

I'd like to thank Gino Torretta, the Heisman Trophy-winning quarterback from The University of Miami (AKA The "U"). We worked together on the set of *Stuck On You* back in 2003, and meeting you again off the set in the neighborhood was awesome. As an ultimate 'Canes fan, meeting you and Sambuca truly let me know that I was definitely at the right place at the right time.

I'd like to thank Derek Fonseca, owner of *South Beach Bartending School* in South Beach (where else?). Thanks also 'Dre and Ivon. Derek gave me the opportunity to make some serious cash and have an absolute blast doing it. It was a great 9 to 5 (9pm to 5am!) I will never forget those all night bartending binges in Miami. Everyone loves a "Sex on South Beach!"

I'd like to thank all the jobs that I've had along the way: from my first job at BellSouth (which was disastrous), to my most recent job at Consolidated Edison (that inspired many of the ideas in this book.) I should also shout out to all the jobs I've had in between, simply because most of them taught me that I was better suited to something other than 'just working a job.'

I'd like to thank the neighborhoods of Frayser (a.k.a. "the Bay") in the city of Memphis, Coral Gables in Miami, and the Bronx in New York City for their molding influence on me. All of the difficult situations I encountered and overcame in these places established my will to succeed.

I know I'm probably leaving out a lot of people, and you may be thinking, "How could Kolie possibly forget me?" If I did forget to mention you, please charge it to my head and not my heart. Oh, let me thank Brother Leroy Clarke for that one. What's up NUPE!

And last, but certainly not least, I'd like to thank you, the reader. Of all the books to read, you chose to pick up mine. For that, I am truly thankful. I honestly believe the principles in this book will help you to create that personal power in your life, allowing you to shine bright for the whole world to see!

*The Goat that Fell into the Ditch

Once upon a time a goat was running through a field, when all of a sudden, it fell into a ditch. The goat had strong legs and had fallen before, but this was an unusually deep ditch; a ditch too deep for him to simply jump out of. In addition, the walls of the ditch were too steep for the goat to climb. The goat was trapped.

As the goat stood in the bottom of the ditch, it became depressed as it could see no way to get out of the ditch. Then it noticed movement at the top of the ditch. It was a man.

The goat thought, "Great. Maybe this man can help me out of this ditch!" So the goat began to kick and scream to get the attention of the man. But then the goat noticed that the man had a shovel. The man began to shovel dirt back into the ditch.

The goat was confused and frightened as dirt, rocks, and debris rained down on top of him. Now, in addition to being trapped at the bottom of a deep ditch, the goat began to become buried. Dirt covered the goat half-way up its legs, and it started to panic because the man began piling the dirt into the ditch faster and faster. Dirt and rocks came down on top of the goat and covered more and more of the goat.

As more and more dirt was filling the hole, the goat was covered up to his neck. The goat was being buried alive! Just as it thought all hope was lost, the goat began to move and jump around. It freed itself enough to *jump on top of the dirt that had been dumped on top of it.* As it jumped and moved, the dirt that was being piled on top of it became its platform.

As the man kept shoveling dirt into the ditch, and the goat fought to keep the dirt under it instead of on top of it, it noticed that the ditch did not seem so deep anymore. The goat was now standing on top of a pile of dirt that would have completely buried it if the goat had not fought against it. The dirt, which the goat at first perceived as its problem, was now filling the ditch and was becoming its solution. The dirt was filling the ditch and bringing the goat, which was fighting to stay on top of the dirt, closer to the surface.

The goat suddenly understood what was happening: he was nearly saved! As more and more dirt was piled into the ditch by the man, the more excited the goat became. The goat could now look up and see the top of the ditch, which was now only a few feet above. It was almost there, but the goat's legs were getting tired. More and more dirt piled in, and the goat had to fight harder and harder to stay on top of the dirt.

Finally, at last, the goat was within jumping distance of the top of the ditch. With one last strong jump, the goat freed itself of the dirt, and emerged triumphantly from the ditch!

After the goat was out of the ditch, it looked around and saw the man standing there with a shovel at his side. The goat asked the man (Yes, I know animals don't talk. But this is a fable, so just play along), "Were you trying to bury me alive, or were you trying to help me out?"

The man replied, "What I was trying to do to you is not important. What *is* important is how you *responded* to what I was doing to you."

HOW TO READ THIS BOOK

I'm sure we all remember playing the "analogies game" in grade school. Two simple examples are as follows:

Finger is to Hand **as** Toe is to ___Foot___.

Car is to Street **as** Boat is to ___River___.

Learning an analogy is almost like learning to speak a foreign language, although much simpler. If I were to say "thank you!" in English, it means the same as "gracias!" in Spanish, or "merci!" in French. Three different words, but they all have the same meaning.

Well, that's exactly how it is going to work here, except with electrical principles and life principles. This is a book of analogies, and I am going to show you how to string these electrical analogies together, and then organize them to form a complete thought: ***SUCCESS!!!***

Traditionally, going from a word in one language to a word in another language with the same meaning is known as a "translation". In this book we will be going from one word in one language to another word *within* the same language. We will call this transition an "*Interlation™.*"

I am going to show you how to organize and plan your life using the concept of the electric circuit. The manner in which the parts of the electric circuit are organized determines, to a great extent, whether the light will come on or not... in both the circuit, and also in your life.

By the end of this book, when you hear "voltage", you should automatically think "motivation". When you hear "electric current" you should automatically think "action". When you hear power, you should automatically think... (See Chapter Six).

The book is divided into two sections. The first section of principles identifies the four components (Chapters One through Four) of a basic electric circuit and demonstrates how they are organized (Chapter Five) to create electric power (Chapter Six). Ironically (or perhaps not so ironically), there are four components in life that must also be organized to create personal power.

The second section of principles (Chapters Seven through Thirteen) helps you to incorporate these principles into the basic electric circuit created in the first section, and demonstrates how they increase or decrease power.

Additionally, each chapter is represented by a symbol. Each symbol represents both the electrical principle, as well as its analogous life principle. The chapters end with the symbol being shown in relation to the previous chapter symbols. Look closely at these symbols because they are very helpful in understanding both the electrical principles, and the process of creating, increasing, and decreasing power—all of which will help you to live a more successful and fulfilling life.

Taking a cue from my good friend & mentor Monroe Mann, I encourage you to take this book with you everywhere. Even if you have only five minutes of free time during a hectic day, this book can be very beneficial. Simply glancing at some of the quotes in the *"Boost Your Frequency!!!"* section in the back of the book may prove to be the most inspiring five minutes of your day!

In addition, read this book frequently and mark it up. Yes, write in this book. Make notes in this book. Highlight the topics in this book that hit *you*. Make this book *yours*. This is not some $2000 set of encyclopedias that just sits up on the bookshelf to look pretty. The only good book is a *used* book.

Finally, read this book as though I am talking to only you. You and I are simply sitting down and having a conversation. These principles are not as complicated as you may think. I wrote this book specifically for YOU, and I hope it helps you greatly on your life journey.

PART I

CHAPTER 1

LIGHT – Your Realized Dream

ANALOGY: *Turning on the Light is to the Electric Circuit as Realizing your Dream is to your Life Circuit*

INTERLATION: "Turning on the Light"~"Realizing Your Dream"

Let your light shine. Shine within so that it can shine on someone else. Let your light shine.

- Oprah Winfrey

A leaf, a sun-beam, a landscape, the ocean, make an analogous impression on the mind. What is common to them all, -- that perfectness and harmony, is beauty. The standard of beauty is the entire circuit of natural forms,--

-Ralph Waldo Emerson

So *what* exactly is it that we are trying to do here? *What* do we want? *What* is the desired outcome? **Turning on the Light is the analogy we will use for the realization of a dream.** It's actually pretty simple. If the light is off, your dream has not yet been realized. You've got some work to do. If the light is on, then your dream has been made a reality!

Figure 1.1 – Light OFF (Unrealized Dream) and Light ON (Realized Dream)

In the first picture, the light is off. In the second picture, the light is on. Pretty simple, right?

Keep this concept in mind as you read through this book. You must always know what your light is! Otherwise, how do you know whether it is on or off?

Although nothing is more important than knowing exactly what specific light it is that you want to turn on, the actual choice of the light really doesn't matter. Who cares what it is as long as *you* know what it is and *you* choose it and it is a light that *you* want to turn on. What I'm saying here is that *you* must know the dream you want to realize and *you* must choose it—no one else can do it for you.

I stress this point for two reasons: 1) Many, many people have no light at all. They have no clue what they want to do with their life. They have no dream to realize. Can you turn on a light that doesn't exist? No, you cannot. 2) Most people are too easily influenced into turning on a light that is not truly theirs. How many people do you know who are doing something they absolutely hate because they were influenced by people around them to do it? The problem with turning on a light chosen by someone else is that the 'beautiful glow' doesn't actually belong to you; it belongs to the person who influenced you to do it.

There is a saying that opportunity is like a gold mine. If you don't pick up the shovel and work your own mine, then you will be living in someone else's shaft.

Never let anyone choose your light! Choosing your light is both your responsibility and your right. Again, the light you choose is not important. What is important is that you do in fact choose one, because if you don't ever have a light to turn on, your light will effectively always be "off", and you will spend your life working to turn on someone else's light (i.e., you will be living in their shadow.)

Think for a second about how many different types of lights are out there. There are regular light bulbs, fluorescent lights, flood lights, street lights, head lights, Christmas lights, and many more. Even within the category of Christmas lights, the light can be green, red, blue, white, orange, flashing, bright, dim, etc. I mention this because the type of light is not important. The important idea is that each light, no matter the type, is either "on" or "off". The same holds true in life. Different people will have different types of dreams. However, each of these dreams either has or has not been realized. Either the light is "on" or "off". That is the focus of this chapter.

Don't Get Overwhelmed

In choosing your light, do not get bogged down in the beginning about exactly *how* you are going to turn it on. The larger your light (and thus the bigger your dream), the less clear it will be in the beginning as to *how* you will actually turn it on. At this point, the "how" is not important because help from other people and better information will become available to you as you move forward.

Focus on exactly *what* light you want to turn on, and *why* you want to turn that light on. Keep this important concept in mind.

Here is an example to illustrate. Fundamentally, turning on 40,000 lights in Yankee Stadium is very similar to turning on a single light in your kitchen. The only difference is that you have a much greater amount of light to turn on. Therefore, you require a much greater amount of power (which we will discuss in Chapter Six) to turn them all on. But the *process* of turning on light, whether it is 40,000 lights in Yankee Stadium or one light in your kitchen is the same: there is either enough power to turn on the light(s), or not.

Now, paralleling the electrical principles, turning on a single light can be seen as the equivalent of realizing a small dream, perhaps learning your alphabet backwards. On the other hand, turning on the 40,000 lights in Yankee Stadium may be seen as a huge dream, similar to becoming the first person to walk on the surface of Mars. Just as it will take more power to turn on those 40,000 lights than the one single light, it will similarly take more power to realize the huge dream of becoming the first person to walk on the surface of Mars than to realize the small dream of learning your alphabet backwards.

I will show you how to create sufficient power to realize your dreams in Chapter Six. But for now, the key is to understand that no matter what the dream—and no matter whether it is big or small—realizing your dream is analogous to turning on the light. This is the first and most important concept in the book, and it bears repeating: realizing your dream is analogous to turning on the light. Got it?

Before we move on, you may as well know something from the start. Large lights also emit large amounts of heat. The lights shine very hot and bright! If you turn on a large light (realize a big dream), you are going to create a lot of heat and you will outshine many people around you who are still dim. This has a tendency to make some people uncomfortable. The easy thing for them to do is to try to dim your light so you do not outshine them and make them uncomfortable with the heat you put out. Watch out for these people, otherwise known as "haters".

What they should actually be doing is working to turn their light on as well, but that requires effort and dedication. It is much easier to try to dim your light. A smart person whose light is dim will seek out a person whose light is on, and try to find out how they managed to turn it on. And when they come to you (because your light is shining), that's when you say, "The first thing you need to do is go out and buy *Electric Living: The Powerful Life!* by Kolie E. Crutcher III."

But back to the lesson: A light does not just turn on by itself. This is evident by the countless hours of work put in by inventors such as Thomas Edison and Lewis Latimer while trying to perfect the light bulb. Thomas Edison and his staff tried over *10,000* combinations before they found one that worked.

Do not be confused, though. Turning on your light is not a matter of chance. *Specific* parts must work together and be organized in a deliberate plan. It's exactly the same with life. Your dreams will not just "turn on" by themselves. You must bring together certain parts and organize them in a specific and methodical way. Edison *knew* exactly the light he wanted to turn on. Ironically, the light he wanted to turn on (his actual dream) was…literally just that—the light bulb. ☺

Do you now better understand how the electrical principle of **TURNING ON THE LIGHT** is analogous to the life principle of **REALIZING YOUR DREAMS**?

If so, it's now time to move on to Chapter 2, and to introduce you to the electric circuit. And after reading Chapters 2, 3, 4, & 5, you will never look at an electric circuit in quite the same way again.

CHAPTER KEYS

- The light you want to turn on is the dream you want to realize. Equate the two.

- Choosing your light is your responsibility and is your right. No one can do it for you.

- The light you choose is not important. What *is* important is that you choose one.

- Do not get overwhelmed with *how* you are going to turn on your light. The bigger your dream, the less likely you are in the beginning to know *how* you are going to realize it.

The Life Circuit™

LIGHT ~ *DREAM*

CHAPTER 2

ELECTRIC CURRENT – You *in Action!*

ANALOGY: *Electric Current is to the Electric Circuit as your Actions are to your Life Circuit*

INTERLATION: "Electric Current" ~ "Action"

Some people want it to happen, some wish it would happen, others make it happen.

- Michael Jordan

Do not wait to strike until the iron is hot; but make it hot by striking.
- William B. Sprague

No Rules. No Excuses. No Regrets.®

- Monroe Mann

Electric current is defined as *the continuous directed flow of electrons from atom to atom.* Depending on your knowledge of science, you may ask, "What is an electron?" Without getting overly technical, an electron is the part of the atom that is free to move.

Okay, so what's an atom?

The Basic Atom

Everything on this earth is made up of very, very, very small parts called atoms. Because atoms are so small, we cannot see them individually. But if we could, they would look something like this:

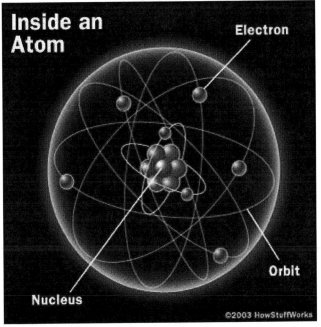

Courtesy of HowStuffWorks.com

Figure 2.1 – Basic Atom

Normally, electrons simply move around in an orbit around the nucleus inside their atom:

However, under certain conditions, electrons will flow continuously from atom to atom:

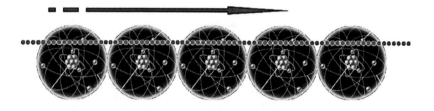

This continuous flow is called electric current.

In order to better visualize the difference, it may be helpful to think about how the water in a pond differs from the water flowing in a river.

The water in a pond does not go anywhere. It is not in motion flowing toward some other destination. On the other hand, the water in a river is constantly in motion, never in the same place as time passes. This is why there is a much better chance of finding algae and stagnant water in a pond than in a river: the water in a river is constantly in motion, and therefore, algae and stagnation do not have time to form because river water is always in motion, flowing on to bigger and better things.

How then does this principle of electric current relate to our personal lives? Well, let's consider two hypothetical people, both different. One of them is a person with no definite direction who moves through life by wandering around in his same common environment day after day. He stays around the same common people and does the same common things in the same common places. This is the motion of a plain old electron. How boring!

Now think about someone you know who moves in the opposite fashion. This person is always in action. This person moves with quickness and authority, from one place to the next. He goes to new places, meets new people, and sees new things, always on the go! This is the motion of electric current.

So, do you want to simply be an electron, or do you want to be part of the electric current? Remember, electrons and electric current are not the same. The difference lies in the way they move. Electrons "hang out" in relatively the same area inside their atom. But the electric current is the continuous flow of those electrons in a definite direction from atom to atom.

If you want to become a professional basketball player, you have *to go* to practice, you have *to go* to games, you have *to go* watch sports films, and you have *to go* study your competition. These are all things you have *to go and do*. These are all actions—specific actions—that you have to take in order to become a professional basketball player. The going, the doing, and the action in a specific direction for a specific purpose is your electric current.

The best laid plans in the world are of no value unless you can actually go out there, pull the trigger, and make it happen. You must have a plan, yes, but that is not enough; You must plan the work, yes, but you must also *work the plan!* Although planning is important, working the plan is even more important. There is no possible way to complete any plan without definite action being taken. Without action, those plans are useless.

I am not devaluing planning in any way. What I am doing, though, is emphasizing action, because many people use planning and excessive thinking as an excuse for delaying or not taking action.

Bottom line: Get your ass in gear and start moving like electric current! Move fast. Move continuously. Move with direction. Move with authority. Move with purpose. Random, indefinite, half-hearted actions—whether concerning electrons or people—create no power. The mathematical proof of this is given in Chapter Six on power. But for now, understand that people are analogous to electrons. And the types of people who are "go-getters"—constantly moving, flowing, and taking action toward some definite major purpose—are analogous to electric current. Electric current, *not regular electrons*, creates power.

PROACTIVE vs. REACTIVE

WARNING! ALL ACTIONS ARE NOT CREATED EQUAL.

When everything boils down, there are only two types of people—Proactive and Reactive. Proactive people seek and accept responsibility, and tend to initiate their actions based on some desire, want, or need for the future. Reactive people tend to be forced to act to compensate for something that has already taken place because they have not acted quickly enough, do not have a plan, or do not have a dream strong enough to cause them to move on their own. The Universal Law of Motion is simple: Either move (Proactive) or be moved (Reactive).

Proactive people get rich. Reactive people get what is left over after the Proactive people are done.

It's really that simple. The universe seems to be in favor of those who initiate action rather than those who must be told or forced to take action.

Think about that deeply as we go over the key points for this chapter.

CHAPTER KEYS

- The continuous directed flow of electrons from atom to atom is known as Electric Current.

- You are an electron. As such, you create no power unless you are continuously flowing in a definite direction, i.e. you should seek to live like an electric current.

- Proactive people get rich. Reactive people get whatever is left over after the Proactive people are done.

The Life Circuit™

LIGHT ~ *DREAM*

Most recent analogy appears in box

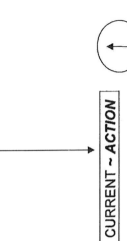

CURRENT ~ *ACTION*

CHAPTER 3

VOLTAGE SOURCE – Your Source of Motivation

ANALOGY: *Voltage is to the Electric Circuit as Motivation is to your Life Circuit*

INTERLATION: "Voltage" ~ "Motivation"

Motivation is a fire from within. If someone else tries to light that fire under you, chances are it will burn very briefly.

- Stephen R. Covey

Okay, from Chapter Two we know that "under certain conditions" electrons will move continuously in a directed flow from atom to atom. This is electric current.

We also know that people are like electrons, and people who are "go-getters" are like electric current because they are continuously flowing and taking action toward some definite major goal or dream.

But *why* do people get moving toward the attainment of a goal or dream? The "why" is the topic of this chapter. By discovering why electrons take definite action, we can discover why people take definite action, and apply the principle to our lives.

The question is, "What exactly pushes or motivates electrons to flow from atom to atom?" I mean, they don't just move on their own. The push or source of motivation of electric current is known as *voltage*. In some text books, this voltage is even referred to as electro*motive* force. Look again at this word.

In fact, let's look closely and break it down together. Electromotive force is the force that *moves* electrons. It is the source of motivation for those electrons to become electric current.

What is your source of motivation? What is your voltage? Going back to electric current, we can see that if there is no voltage pushing on the electrons then there is no flow. There may be a few electrons moving around inside their atom, but there is no continuous directed flow, and thus no electric current.

So how does the principle of the voltage source apply to you and your life? Simple: electric current will not flow without a source of voltage—a source of motivation.

This is also true of people as well. Think about it. You, me, and every other person on the face of this earth needs a motivation before we will take any action of any kind. This is true even of the simplest actions. Most of the time, we are not even aware of what our motivation to take a certain action may be, but it is *always* there. Let's take this very moment. What is your action? You are reading this book. See if you can figure out what the source of motivation is that caused you to take the direct action of reading this book. Your motivation may be that you want to learn about the basics of electricity. Your motivation may be that you want to create more power in your personal life or business. Your motivation may be that you want to read a certain number of self-help books by the end of the year. The point is that no one moves in any direction or takes any action without having sufficient motivation to do so.

Let's look at a few examples:

Source of Motivation	**Resulting Action Taken**
HUNGER	YOU EAT
ITCH	YOU SCRATCH
GOOD CREDIT	YOU PAY YOUR BILLS ON TIME

Figure 3.1 – Sources of Motivation and Corresponding Actions Taken

If you are not doing something, then it is simply because your motivation has not reached a high enough level to push you into action. Your motivation is not strong enough to cause you to hustle and flow. I've been there. I know.

And here's the solution: stop making excuses and get real with yourself. Every task, job, or function requires a certain amount of motivation in order to be accomplished. Some require more motivation than others. For instance, it takes more motivation to become a professional basketball player than it does to watch Jerry

Springer and Judge Judy all day long. You might also notice that if you are doing one thing, it consequently then means you are *not* doing something else. Think about it. There are only 24 hours in a day. Not only does your motivation cause your actions, but *your actions reveal your motivation* (Actions do speak louder than words.) If I want to know what you want—what you *really* want—all I have to do is look at what you *do*. And acting in a manner that is the opposite of what you want to do makes a person very unhappy. This contradiction between what you want and your corresponding actions will not last long. Sooner or later, your actions must fall in line with what you truly want. And it could be very painful to discover that what you really wanted all along… was failure.

Do you now understand how the electrical principle of VOLTAGE parallels the life principle of MOTIVATION?

Is Your Voltage Source The Wall Outlet or A Duracell?

It is very important to know your source of voltage because you must know how to get more of it when needed, and also how often you will need to replenish it.

If you have a device that is battery powered, such as a flashlight, it would be foolish to not have extra batteries around. On the other hand, it would also be foolish to expect to use a device that plugs into the wall outlet, such as a hair dryer, and neglect to pay the electric company.

There are only two ways to get voltage: through a cordless source, or a wired source. If you need voltage for a flashlight, you can get it from a Duracell battery. If your flashlight begins to dim or goes out, you must know to get more Duracell batteries. If you need voltage to power a hair dryer, you must know to get it from the electric company (wall outlet). If your hair dryer does not work (and there is nothing wrong with the dryer itself), you must know to pay the electric company. You need to understand the difference between the two sources of voltage. For instance, if your flashlight does not work and you are trying to contact the electric company, you are in trouble.

In the same manner, you as an achiever must know what motivates you. You must know your source of voltage. Personally, pictures of Lamborghinis and Ferraris motivate me. Certain music motivates me. If at any particular time, I am feeling lazy and don't want to take action towards my goals, I turn on my Ipod or pull out my Lamborghini pictures. That motivates me. It would be foolish of me to bake a cake or plant some flowers when I feel unmotivated. Those things do not motivate me. To each his own. Your motivation is *yours*. Use it!

Motivation is also analogous to the voltage source because no source of voltage lasts indefinitely without re-charge. Just as batteries will fade out and die, your motivation will fade out and die if you do not deliberately re-motivate and re-charge yourself daily. According to Zig Ziglar, "People often say that motivation doesn't last. Well, neither does bathing. That's why we recommend it daily." And Andrew Carnegie once said that people who are unable to motivate themselves must be content with mediocrity, no matter how impressive their other talents.

Just as the amount of voltage that pushes electric current can vary, some people have stronger motivation than others. You guessed it. People with stronger motivations push themselves more and take more action than people with weaker motivations. A person who has a high level of motivation is always moving, flowing, and taking action toward his or her goal. That person is known as a self-starter or "go-getter."

CHAPTER KEYS

- The Voltage Source in an Electric Circuit is what "pushes" or motivates the Electrical Current.

- You must know what your motivation is. Only then can you continuously re-motivate or "re-charge" yourself.

<u>Note</u>: This is *the most* important principle in this book after the idea of light as your dream. It is *absolutely essential* that you understand this relationship between your motivation and success. No one can give you your motivation. Each individual must know what it is that causes him to act. Without a continuous source of motivation in your life, you will not have sufficient power to achieve any dream. I call it your voltage; Monroe Mann calls it your 'Why'; it doesn't matter what you call it, but what does matter is that you understand that without it, you will have a lot of difficulty reaching your dreams.

The Life Circuit™

LIGHT ~ *DREAM*

CURRENT ~ *ACTION*

VOLTAGE ~ *MOTIVATION*

CHAPTER 4

RESISTANCE – Fear

ANALOGY: *Resistance is to the Electric Circuit is as Fear is to your Life Circuit*

INTERLATION – "Resistance" ~ "Fear"

You gain strength, courage, and confidence by every experience in which you really stop to look fear in the face. You are able to say to yourself, "I lived through this horror. I can take the next thing that comes along." ...You must do the thing you think you cannot do.
- Eleanor Roosevelt

You can't fly a kite unless you go against the wind and have a weight to keep it from turning a somersault. The same with man. No man will succeed unless he is ready to face and overcome difficulties and is prepared to assume responsibilities.
- William J.H. Boetcker

...And even when my days were the bluest, never ran from adversity, instead I ran to it. Fear ain't in the heart of me, I learned just do it. You get courage from your fears—right after you go through it.
- T.I.

The basic idea behind the light bulb is simple. Electric current flows through the filament of the light bulb, which glows from the heat, thus providing light.

The filament is a very long and incredibly thin material (usually tungsten) within the bulb. If you were to look closely at a light bulb while it is off, the filament is the thin coil in the center. In a typical 60-watt light bulb, the filament is about 6 ½ feet long, but only 1/100 of an inch thick. Because the filament is so long and thin, it offers a large *resistance* to the flow of the electric current.

Electric resistance is the total measure of the opposition faced by electric current as it is being pushed by the voltage source. There are several factors that influence the resistance of an object. Length, width, composition, and even temperature influence the amount of resistance the object sets up to oppose the flow of electric current.

Resistance is the equivalent of fear in your life for the following reason: *Just as electric resistance sets up opposition to the flow of electric current, fear sets up opposition to the flow of your actions.* Fear makes it more difficult for us to act. Think about it.

What would you attempt if you had no fear? Your actions would be limitless. Fear stops more people faster than anything else. Fear has been the downfall of more people than any single reason ever. More specifically, what has caused the downfall is the fact that these people allowed fear to keep them from taking definite action toward their dreams, without even realizing that fear—like resistors in a circuit—is necessary for success.

That fact is rather interesting given that when we are born into this world, we have only two fears: the fear of loud noises and the fear of heights. All other fears are acquired as we grow and develop.

While having fear is natural—and while some fears are necessary for us to survive—when fear prevents us from acting, from doing the necessary things to achieve our goals and become successful, that is not good. I have a saying that I keep near to me at all times. It is this: "This *thing* that you are afraid of is not the issue. The issue is: can you *act despite* the fear?" A corollary to this is given in *The Magic of Thinking Big* by Dr. David Schwartz. He gives the cure for fear. It is simple. The cure for fear is *action*. Action cures fear. Simple, right? Yes it really is. But it is not easy.

You see, many times we are not even aware that we are being paralyzed by fear. And this is something we must learn how to recognize before we can make the adjustment.

Have you ever heard the term, "like a deer frozen in the headlights?" Many of our fears are deep within our subconscious, where they are difficult to find, and even more difficult to deal with. The question is, how would you live your life if you knew you could not fail? And from there, that is where you should begin taking action.

The people in this world who move up and achieve are those who courageously attack that which they fear. They stare their fear right in the face and move toward it as opposed to stopping and chickening out. That does not mean they are not scared. They may actually be terrified. But they still manage to take action!

When you face your fear head on, you actually become more familiar with it, and you become less and less afraid. We fear what we do not know or understand. And by facing our fears we actually become more familiar with them and they become less frightening.

No matter how bad the circumstances *appear*, never stop moving. Never allow fear to stop your flow. Anything that does not continue to move will die. When you flip a light switch "off", you have stopped the flow of electric current through the light bulb's highly resistant filament. No movement through resistance means no light. Fear is nothing more than a test to see how badly you really want something you said you wanted. That's it.

Now that we have introduced and defined electric current, voltage, and resistance, we can introduce the very special equation that relates them to one another. Don't worry; it's easy to understand.

$$I = V / R$$

Where,

I = ELECTRIC CURRENT
V = VOLTAGE
R = RESISTANCE

This equation is known as *Ohm's Law*. In everyday language, this equation means the amount of electric current that flows is simply the ratio of the voltage to the resistance.

Look at how Ohm's Law fits right in line with the equivalent law that governs our actions in life:

ACTION = MOTIVATION / FEAR

Is this not true in our lives? Isn't our level of action dictated by our level of motivation divided by our level of fear? I am going to drive this point home and make it perfectly clear by using what I call the "Magic Action Chart". Check it out. You are going to like this!

First, at any given point in time, we can rank how much we are doing in our life, what we are moving towards, and our level of action on a scale from one to ten, with one being the lowest level and ten being the highest level.

ACTION LEVEL

The same holds true for our level of motivation and our level of fear. Each can be ranked on a scale from one to ten.

MOTIVATION LEVEL

FEAR LEVEL

What may not be so obvious from this introduction is that the level of your action is determined by your level of motivation and your level of fear.

I will give you three quick examples that explain graphically why some people take a great deal of action, why some people take very little action, and why some people take no action at all.

Example No. 1 - Tim

Let's say that Tim is a very motivated person who has very little fear. His motivation and fear rankings may look something like this:

MOTIVATION LEVEL - 8

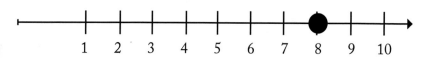

FEAR LEVEL - 1

According to the equation:

ACTION = MOTIVATION / FEAR

Therefore, in Tim's case we have:

ACTION = 8 / 1

Therefore, Tim's level of action is 8.

ACTION LEVEL

It makes sense that Tim is ranked on the high end of the action scale because his motivation is so much greater than his fear.

Example No. 2 – Roger

Let's say that Roger is a somewhat motivated person but he lives in a state of moderate fear. His motivation and fear rankings may look something like this:

MOTIVATION LEVEL - 5

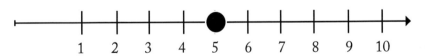

FEAR LEVEL - 5

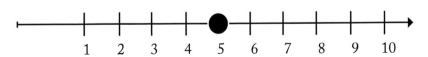

According to the equation:
ACTION = MOTIVATION / FEAR

Therefore, in Roger's case we have:

$$ACTION = 5 / 5$$

And Roger's level of action is 1.

ACTION LEVEL - 1

It makes sense that Roger is ranked on the low end of the action scale because his motivation and fear are equal. They cancel each other out and the result is very little action.

Example No. 3 - Andy

Let's say that Andy is a highly unmotivated person who lives in constant fear. His motivation and fear rankings may look something like this:

MOTIVATION LEVEL - 2

FEAR LEVEL - 8

According to the equation:

$$ACTION = MOTIVATION / FEAR$$

Therefore, in Andy's case we have:

$$ACTION = 2 / 8$$

And Andy's level of action is 0.25

ACTION LEVEL – 0.25

Because Andy's level of action is below 1, Andy does not act *at all.* It makes sense that Andy is ranked below 1 on the action scale because his motivation is far less than his fear. Fear wins out and the result is no action.

So summing up:

- **As the voltage to resistance ratio gets larger – MORE ELECTRIC CURRENT**
  As your motivation to fear ratio gets closer to 10 – you ACT MORE

- **As the voltage to resistance ratio gets smaller – LESS ELECTRIC CURRENT**
  As your motivation to fear ratio gets closer to 1 – you ACT LESS

- **If the voltage to resistance ratio drops below 1 – Negligible ELECTRIC CURRENT**
  If your motivation to fear ratio drops below 1 – you will NOT ACT

Do not be led to believe that life's ideal condition would be one in which you never encountered any fear. It may be easy to fall into the trap of thinking less fear automatically means more action, but remember that the level of action depends on the **ratio** of motivation to fear, not fear as a pure number by itself.

In fact, as my good buddy Monroe Mann pointed out to me while helping me edit the book, there is another situation that a lot of people find themselves in that I didn't mention. If your motivation is maxed out at 10, but your fear level is also a 10, as we now know, you will not take much action. You may dupe yourself into believing that you're a really motivated person, and you very well may be, but because your fear level is also high—and that can be fear of failure *or* fear of success—then your high level of motivation really means very little—because the fear is far more powerful. In this situation, your only hope for action is not an increase in motivation, but rather, a decrease in fear.

IMPORTANT SIDE NOTE: Ohm's Law can be manipulated to read:

$$V = I{*}R$$

This is the same equation from earlier in the chapter, solved for Voltage. It reads, "Voltage = Electric Current times Resistance." I mention this for only one reason. Take a look at the word the equation spells: VIR. It is very interesting to note that in Latin, "VIR" means "Man".

The equation for motivation, therefore, is related to Man. Man, you see, is meant to be motivated!

CHAPTER KEYS

- Fear's goal is to prevent us from moving toward our goals and dreams.

- We stop moving because we are unaware of what is on the other side of fear: tremendous opportunity.

- Action is the cure for fear.

The Life Circuit™

LIGHT ~ *DREAM*

RESISTANCE ~ *FEAR*

CURRENT ~ *ACTION*

VOLTAGE ~ *MOTIVATION*

CHAPTER 5

ELECTRIC CIRCUIT – The Planned Path to Your Realized Dream

ANALOGY: *The Electric Circuit is analogous to the Planned Path of your Life leading to your Dream*

INTERLATION: "Electric Circuit" ~ "Planned Path"

If you fail to plan, plan to fail.

- Author Unknown

Plan the work and work the plan.

- The First of Consolidated Edison's Six Core Principles of *The Way We Work*

An electric circuit is a complete and planned path through which electric current flows via conductors (Chapter Eight), and circuit elements such as resistors, capacitors, and inductors (Chapters Four, Eleven, and Twelve respectively).

The electric circuit is the place where the electrical phenomena we have been discussing take place—the "stage", if you will. In the same fashion, our life can be thought of as the "stage" on which events take place.

In Chapter Two we learned that electrons will, if pushed, flow at a definite rate through the electric circuit. This flow is known as electric current. In Chapter Three, we learned that the push that causes the electric current is called voltage. In this chapter, we switch our attention to the path of the electric current flow and where that path leads. We must not overlook the fact that electric current will not flow without a path.

How do we ensure that our electric current travels toward our resistance in our light bulb and creates power? *We have to plan out the path of the electric current path with the end result in mind.* A plan provides the creator of the plan a great deal of control over the outcome of the situation he has at hand.

When we are building an electric circuit, an important part of the process is carefully choosing the type and position of conducting wires. The position of the conducting wires determines where the electric current can flow. This is very important!

If I want electric current to flow from Point A to Point B, I must provide a planned path from Point A to Point B through which that electric current can flow. This path keeps the current "on track", so that it does not flow to unintended places. I have a great deal of control over the flow of electric current simply by planning out the current's path. Without a path, there is no way to control the flow of current.

This is also the reason having a plan for your life is so crucial. Without a plan, there is no way to know where you are going, much less control where you are going. Just as with electricity, action that is not planned out will lead either nowhere or to somewhere you do not desire to be.

Here's an example. Let's say that we want to travel from New York City (Point A) to Miami Beach (Point B) for Memorial Day weekend. Whether we succeed in getting down to Miami Beach at the time we desire depends in large part on the plan we develop in order to get there. Do we drive? Do we fly? Do we have money? Do we have identification? The more detailed our plans, the better chance we have of actually getting to Point B—Miami Beach—when we want.

A plan, then, is the path you (electrons on the move) travel down as you move toward turning on your light. You need to know where you are going and you need to know exactly where you want to end up.

When you are planning your path, make sure that it is your unique path. Two people can get to the same location without traveling down the same road. So it is important to realize that your path may be slightly or even vastly different than those who are around you, even if you are going to the same final destination. This leads us to series and parallel circuits.

Series and Parallel Circuits

In life, we always have a choice in deciding which actions we take and the path we move along. Similarly, the electric current in our circuit can flow through different paths, depending on how the circuit is planned.

An electric circuit, planned in such a way that the electric current has only one path it can travel, is known as a *series* circuit. An electric circuit that is planned so that the electric current can travel down multiple paths is known as a *parallel* circuit. Here is an example of each:

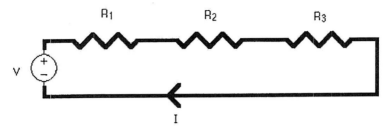

Figure 5.1 – Flow of Electric Current (I) through Series circuit

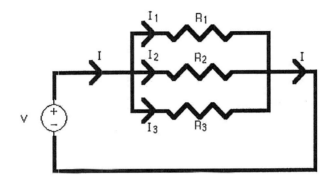

Figure 5.2 – Flow of Electric Current (I, I_1, I_2 and I_3) through Parallel circuit

You must plan your path accordingly, depending on what you desire out of your life.

Generally speaking, there are two paths you can take in your life. Those two paths are the "secure" path, which ultimately leads to failure, and the "path less traveled", which leads to freedom. There may appear to be many different paths, however all paths break down to one of these two. In the same fashion, any circuit can be broken down to an equivalent series or parallel circuit.

If you want to live the "secure" life, you will plan your life as a parallel circuit, giving yourself multiple options, and something to "fall back on." You will choose a "good" job with "good" benefits and a 401k. But understand that the worst enemy of "the best" is "the good".

On the flip side, planning your life as a series circuit qualifies you for the greatness of your destiny. There is something magical about putting all your thoughts and actions behind the attainment of a single goal, without allowing yourself the option to do something else "just in case it doesn't work out." You develop *focus*. And when you have focus, you are capable of infinitely more than when your thoughts and actions are spread out over multiple interests. Focus is the secret of both the magnifying glass and the laser.

Think about powerful and successful people. If I say, "Michael Jordan", what is the *one* word that comes to mind first? Basketball. If I say "Muhammad Ali", what do you think? Boxing. "Bill Gates"? Microsoft. The point is that successful people choose *one* field in which to specialize in, then put all of their energies behind their goal. For them, the issue is not whether they will succeed, but *when* they will succeed.

John H. Johnson, the founder of Johnson Publishing Company, Inc., summed it up best when he once stated, "The key to my success has been to give up everything for my dream." Johnson even took his focus a step further. "I decided once and for all that I was going to make it or die."

<u>Burning Your Bridges</u>

Once upon a time, a great warrior faced a decision to ensure his success on the battlefield. He was sending his army against a powerful enemy whose men outnumbered his own. He loaded his soldiers into ships and sailed to the enemy's land. Upon arrival, he unloaded his troops and equipment, and burned the ships that had carried them. He then addressed his men before the first battle, "You see the boats going up in smoke. That means that we cannot leave these shores alive unless we win! We now have no choice—*we win—or we perish!*

They won.

CHAPTER KEYS

- As you travel through life, you must have a planned path. Otherwise, you will not know how to get to your goal, or where you are in relation to your goal.

- In pursuit of your definite major goal, burn all of your bridges behind you, leaving no room for escape. Only then can you create within yourself that laser-like focus necessary to burn through all obstacles.

The Life Circuit™

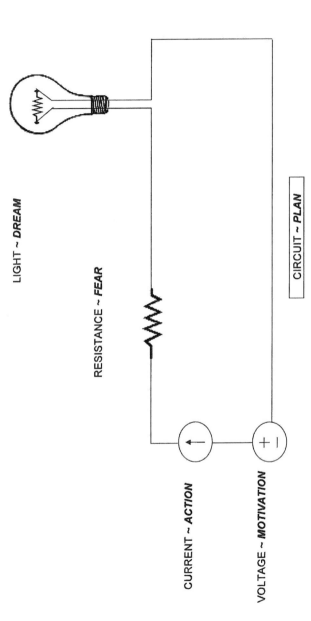

LIGHT ~ *DREAM*

RESISTANCE ~ *FEAR*

CIRCUIT ~ *PLAN*

CURRENT ~ *ACTION*

VOLTAGE ~ *MOTIVATION*

CHAPTER 6

POWER – Action Despite Fear

P

ANALOGY: *Power is to the Electric Circuit as Power is to your Life Circuit*

INTERLATION: "Power" ~ "Power"

This thing that you are afraid of is not the issue. The issue is whether you can still act despite your fear.

- Kolie E. Crutcher III

If there is anything in life we as humans desire more than anything, whether consciously or subconsciously, it is power. With the proper amount of power, you have the ability to acquire just about anything else you want in life. Yet, while everyone talks about power, few people actually know what to do in order to get it.

Electrically speaking, power is the rate at which work is done. It is important to note that power is not just getting work done, but the *rate* at which it gets done. Quick example: Let's say I have 10 screws to drill into a piece of wood. I am given the choice between a regular hand screwdriver and an electric power drill. Which do I choose? I choose the electric power drill because I can drill all 10 screws much more quickly. The power drill has more power than the hand screwdriver.

Another example: your co-worker asks you to fax a copy of a company document to him. When do you do it? You do it when it is convenient for you. No big rush, right? But what if the CEO of the company asks you to fax the same document over to him? When do you do it? NOW! What is the difference? It is the same act, and the same document. The difference lies in the level of power the requester has.

People with power not only get things done, they get things done much faster. The more power they have, the more they get done in less time.

Something very special happens inside a light bulb when the electric current flows through the highly resistive filament. The filament glows white hot and the light bulb shines! It is this process of the electric current flowing through the high resistant filament that creates the power that causes light to be emitted. *Without the high resistance of the filament, no light shines.* Likewise, as human beings, if we are unable to act through our fears, we will not create sufficient power to "shine".

So whether we are talking about electrical power or personal power, power is power. We want to get things done, and we want to get them done fast. When power applies to human interaction, we must consider that there are only 24 hours in a day. You personally can only do so much.

Therefore, increasing power requires the use of other people's time to accomplish your own goals and dreams. People working their entire days at jobs in which they have set no clear personal goals or objectives have no real power because their time is being used to realize the dream of the owner of the company. They are not actively engaged in the generation of their own power. They are actively engaged in creating power for someone else.

The process by which power is created is evident when you look at the life of any person who has attained great fame or fortune. Before they "arrived", they ran into many obstacles, setbacks, and hard times. But they faced their fears and kept on keeping on. The picture painted by the media of their life may be one of no difficulties, but the reality of their journey is quite different. Many of these obstacles would cause the average person to give up. But these "hard knocks" in life are actually great opportunities to generate power, which brings us to the central formula of this book:

$$P = I^2 * R$$

Look at this equation. *This is the equation for power—electrical power as well as personal power.* Read it like this: the amount of power created is equal to the square of the electric current multiplied by the resistance that the electric current faces. Four important lessons are learned from this simple equation.

1) **If you increase either the electric current or the resistance, you will increase the Power.**
2) **$P = (I)^2 * 0 = 0$. In other words, NO POWER IS CREATED when there is no resistance.**

3) $P = (0)^2 * R = 0$: **In other words, NO POWER IS CREATED when there is no electric current.**
4) **The electric current (because it is squared) factors more into how much power is created than the resistance in the circuit.** *In other words: Your Action moving forward is more important than the Fear holding you back.*

The vast majority of people see life's difficulties and obstacles (the things that we fear) as signs to quit or give up. They view them as reasons or evidence that they should stop pursuing their goal. Very, very few people view the difficulties and obstacles of life properly: as opportunity; as a sign that you should keep pushing even harder than before.

Remember: resistance is *essential* to create power. No matter how great the flow of electric current, no power can be created without resistance. If people understood the relationship between resistance and the creation of power, they would see that *fear is the opportunity to create power*, because there can be no power without overcoming the thing that you fear.

Try this: the next time you are in a situation where you are afraid and are tempted to give up, remember the equation for power. First of all, recognize the situation as an opportunity to create power, because there can be no power without moving through fear. Secondly, *take action*, knowing that your efforts *will be squared* as they move toward the creation of that power.

<u>This all leads to an important question: Why do we let fear prevent us from acting?</u>

Very simply, we do not recognize that power is on the other side of our fears, *if we act through them*. Many times, we are going along in life without any problems, and then along comes an obstacle. It is difficult to act despite the obstacle because we only see the obstacle at the time. We don't see what is on the other side. Therefore, the situation seems bleak. That is precisely why we must keep the equation for electric power in mind.

$$P \text{ (Power)} = I^2 \text{ (Electric Current)}^2 * R \text{ (Resistance)}$$

First, recognize the simple analogies between electricity and life.

Power ~ Power
Electric Current ~ Action
Resistance ~ Fear

Now the equation becomes:

Power = (Action)² * Fear

Although this is a very simple equation, it is not easy to implement. We must have the self-discipline to train ourselves not to give up when we are faced with opposition. Only then will taking action in the face of fear become a habit. Just as the electric current must face the electrical resistance to create power in an electric circuit, you must face your personal fears in order to create power in your life.

Bottom line, we must train our minds to see what lies just on the other side of our fears, which is, of course, tremendous opportunity. If we learn to see our fears as merely dues that we must pay in order to create tremendous power, we will keep going despite that fear and take the action we need in order to become successful.

CHAPTER KEYS

- **Power is created when Electric Current flows through an area of resistance.**

- **Being afraid is okay. What is not okay is allowing your fear to prevent you from taking action.**

The Life Circuit™

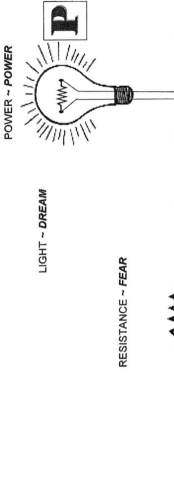

POWER ~ *POWER*

LIGHT ~ *DREAM*

RESISTANCE ~ *FEAR*

CIRCUIT ~ *PLAN*

CURRENT ~ *ACTION*

VOLTAGE ~ *MOTIVATION*

P

PART II

The purpose of the first section of this book was to identify the four components (Chapters One through Four) of a basic electric circuit and demonstrate how they are organized (Chapter Five) to create electric power (Chapter Six).

$$P = I^2 * R$$

Given that we have already created power, our focus now switches to a study of how to keep that power constant, and then, how to increase it.

Obviously, if we increase our resistance (Obstacles and Fears), and also increase our electric current (Actions) in the face of that resistance, we increase our power. Success, as you should have noticed by now, is really a simple formula.

However, there are other important but less obvious principles that—when implemented correctly—maintain and increase power (and when applied incorrectly, severely decrease power). These very principles are addressed in the 2nd section of this book—which you are about to begin reading.

CHAPTER 7

INSULATION – Self-Discipline

ANALOGY: *Insulation is to the Electric Circuit as Self-Discipline is to your Life Circuit*

INTERLATION: "Insulation" ~ "Self-Discipline"

The only way to ensure that you are not under the rule of others is to be under the rule of yourself.

- Kolie E. Crutcher III

No horse gets anywhere until he is harnessed. No steam or gas drives anything until it is confined. No Niagara is ever turned into light and power until it is tunneled. No life ever grows until it is focused, dedicated, disciplined.

- Harry Emerson Fosdick

Okay, so electric current flows in the circuit through conductors as it is being pushed by the voltage source. We understand that already. Now, we need to be certain that this electric current flows to the proper location instead of being diverted to nearby circuits.

There are certain materials known as insulators that do not allow electric current to flow through them. Insulators can be used to cover the outside of a circuit to prevent the electric current from becoming "distracted" towards other nearby circuits. In the same manner, you must protect yourself with self-discipline so that your actions are not distracted away from your dream.

Many people who have cultivated strict self-discipline within their lives are considered to be anti-social, cold, or harsh. This is not the case at all. These people have simply developed the ability to close their mind off to those people, things, and situations that do not lead them to the attainment of their dreams. Their focus on their dreams leaves no room for dilly-dallying and small talk with people who are not focused. They have only one thing in mind, so when it comes to distractions from other people, they can quickly and abruptly say "no", move on, and think nothing of it. Their self-discipline allows them to control their actions and stay on the path to their dreams.

Another reason self-discipline is so important is because it leads to the formation of habits. When an action is repeated over and over again, it becomes part of the natural motor impulses of the body and becomes second nature. However, the action must be repeated many, many times in order to become second nature.

You see, once an action has become a second-nature habit, the thought process is very minimal. You can operate by feel and intuition as opposed to operating from your head. This is the place you want to be because you begin to operate on gut instinct, and on a gut instinct which is nearly always accurate. Any person who has ever been "in the zone" can vouch for this. They may not be able to tell you exactly *how* they do what they do, but they are certainly masters of execution. When asked, they may simply respond, "I don't know.

It just feels right." This is what self-discipline will do. You must be able to insulate yourself from the many distractions of life so that your focus is on your primary goal. If not, your actions will be spread very thin, you will never focus your energy, and the right habits will never become a part of your life.

One of the first full-length plays in which I performed at the American Academy of Dramatic Arts was *Intimate Apparel* by Lynn Nottage. This play was directed by a fabulous gentleman by the name of Gregory Simmons. Greg taught us so much during our rehearsal time, but the phrase I remember more than anything was "practice makes permanent." In other words, *how* you practice determines how you perform. If your practice is sloppy and lacks intensity, your performance will be sloppy and lack intensity. If you practice with specific objectives and a clear point of view, those objectives and that point of view will shine through in your performance. Self-discipline is the key.

If you actually hope to find the motivation and spend the time repeating an action or process *correctly* hundreds—if not thousands—of times until it becomes habit, you need self-discipline. This is true with any of life's endeavors. After all, life really is just a stage. We are all players, whether we realize it or not. The self-disciplined ones will be those in the spotlight:

Self-discipline insulates the aspiring world-class track star from indulging in a diet of potato chips, cheeseburgers, and chocolate cake.

Self-discipline insulates the future doctor from spending his/her nights and weekends partying and drinking.

Self-discipline insulates the up-and-coming entrepreneur from seeking a nine to five job when money gets tight and he/she is being nagged by "friends" and family to get a job.

Self-discipline insulated Orville and Wilbur Wright from the nay-sayers who doubted man would ever tame the skies.

Self-discipline insulated a young Michael Jordan from giving up on basketball when he was cut from his high school squad.

Self-discipline insulated Henry Ford from giving up on his dream to cover the world with automobiles even though many considered him dangerously ignorant.

The key here is that you must protect (i.e. insulate) your mind from all the negativity around you. Every successful person practices strict self-discipline to insulate his mind from the hordes of negative people and situations waiting to pounce on him from every angle.

You need to realize that the people on this earth who regularly set goals, and especially ambitious goals, are few and far between. They are the exception rather than the rule. Therefore they stand out and shine, leaving those in their shadows very uncomfortable.

Misery loves company, and the easiest way for an un-ambitious person to feel comfortable around a person who is a high achiever is to try to bring that high achiever down to his level. Since the un-ambitious person does not know how to rise, he takes great joy in trying to make others fall. It keeps him from looking bad and it allows him to save face.

Self-discipline is the trait that allows individuals to do what they need to do for success in the long run, instead of doing what *feels good* to indulge in at the moment. It causes you to make decisions based on values and principles instead of circumstances or moods.

The Price of Self-Discipline vs. The Price of Regret

Everything has a price—*Everything*. When everything boils down, there are only two prices that you pay for your existence here on earth. One is the price of self-discipline. The other is the price of regret. Self-discipline must be paid up front and in full, but it costs pennies. Regret can be held off and paid later in life, but relative to the cost of self-discipline, regret costs millions of dollars. Here is an example. Let's say that you exercise your self-discipline when it comes to saving and investing your money. If you saved a few dollars a day from the time you are 25, after forty years and through the miracle of compound interest, you would be surprised at the large sum of money it grew to become. Sacrificing a few dollars a day and putting them into an interest-bearing account is something we can all do, and as a matter of fact we would hardly notice it was missing. After a few weeks you become accustomed to saving and it would have by then become a habit. What is the price here? Simply the self-discipline to refrain from spending every dollar you have.

But what if every time you got paid, you spent each and every dollar? What if you never exercised your self-discipline to stash just a small portion of it away? At the end of forty years, you would not have a large sum of money. Instead, you would probably be in considerable debt because you fell into the bad habit of spending all that you had (and more).

People who have lived a life lacking self-discipline look back at their lives with regret. And regret is a heavy price to pay. It slows you down, robs your initiative, and has your head constantly ringing with "what-ifs?"

On the other hand, people who live their lives exercising self-discipline look back on their lives with joy and pride. They have paid up front, in full, for their existence on this earth, and now they move even more swiftly through life, reaping benefits from their earlier investment. By this time, the skill or idea that was developed into a "perfect" habit by self-discipline is now taking care of them.

Again using the earlier example of saving and investing a few dollars a day, the resulting large sum of money is now taking care of the one who took care of the small amount in the beginning. Self-discipline insulated those one dollar bills from being spent on frivolous items, so that they grew, and are now able to be spent on luxurious and extravagant items if desired.

The key here is that without the insulator of self-discipline, those individual dollars would have had no protection. They would have been spent on some other novelty item that would have had no value after it was purchased.

Again, self-discipline leads to habits formed. Once these habits are formed, you will do them automatically. This is the point where you do not have to think about forcing yourself to do what you need to do, because those things have already become automatic. Self-discipline may be difficult at first, but if you are persistent, the reward will be well worth it.

CHAPTER KEYS

- You must insulate your mind and your actions from outside distractions. This insulation is self-discipline.

- People are successful in life because they practice self-discipline which leads to successful habits.

The Life Circuit™

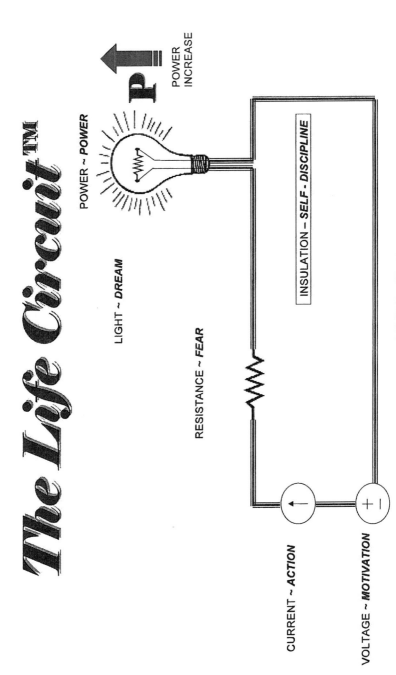

POWER ~ *POWER*

LIGHT ~ *DREAM*

POWER INCREASE

INSULATION – *SELF - DISCIPLINE*

CIRCUIT ~ *PLAN*

RESISTANCE ~ *FEAR*

CURRENT ~ *ACTION*

VOLTAGE ~ *MOTIVATION*

CHAPTER 3

CONDUCTIVITY - Faith

ANALOGY: *Conductivity is to the Electric Circuit as Faith is to your Life Circuit*

INTERLATION: "Conductivity" ~ "Faith"

I believe that the science of chemistry alone almost proves the existence of an Intelligent Creator.

-Thomas A. Edison

Life does not consist mainly, or even largely, of facts and happenings. It consists mainly of the stream of thought that is forever flowing through one's head.

- Mark Twain

Conductivity determines how freely electric current flows in an electric circuit. When electric current meets resistance, less electric current will flow. The opposite of resistivity is conductivity.

From Chapter Three on Voltage, we learned that voltage pushes or motivates the electric current through the electric circuit. For a given level of voltage, more conductivity means more electric current flow.

Because conductivity is the opposite of resistivity, it can be written as 1/Resistance, which is the same as 1/Fear. Another way to say this is "one over Fear". This is a wonderful way to define Faith, "One over Fear". Fear is the *opposite* of Faith. Fear is Faith *in reverse*. They are at two opposite ends of the spectrum, like night and day.

The same is true with conductivity and resistance. The more conductive a material is, the less resistive it is, and vice versa. The more Faith you have, the less fear you have. The more fear you have the less Faith you have.

Consider these questions: Where do ideas come from? How does an inventor or scientist come up with theories and thoughts that others have not thought about? Very simply, there is an *Infinite Intelligence* that is the first cause of everything that each and every mind has *access* to. Most people are not aware that they are constantly in the midst of this Infinite Intelligence and even less aware of how to gain access to it. The genius has learned to gain access to this Infinite Intelligence so that answers to problems that are not readily available become available to him. The information and ideas flow from the great reservoir of Infinite Intelligence into the mind of the genius. The more Faith the person has, the more readily these ideas flow. That's why conductivity is analogous to Faith. More conductivity means more electric current flows.

Again, it is very convenient that Faith is the inverse of fear. You will encounter a great deal of resistance in this world. You need to use it to create power. And as you become accustomed to acting through your fears, those fears will actually turn into Faith. You will no longer fear those things. You will find new fears and challenges to convert into power and Faith through your ability to act.

There will always be fears and challenges to overcome. But as you train yourself to keep pushing you will overcome more and more fear. You will become so full of Faith (conductivity), that you will feel as though you have your own personal direct line to Infinite Intelligence. You become receptive to great plans and ideas and you will begin to get "strokes of genius" on a regular basis.

What you need to understand is that Faith in our lives is the "something" that activates Infinite Intelligence within us. The mind can achieve anything it can *believe*.

Another way to say this is that you do not believe what you *see*; you see what you *believe*. Infinite Intelligence does not give us the capacity to have a vision or dream without also giving us the capacity to bring that vision or dream into reality.

However, we must still have Faith, because Faith is what bridges our human mind to the source of Infinite Intelligence. Once Faith is present in your mind, your mind becomes conductive and the flow from Infinite Intelligence is readily available.

Another way to think about conductivity and how it relates to Faith is to think about a garden hose. Let's say that we have a garden hose with water flowing out. Yes, the water is flowing out from somewhere, but *where it is coming from* has nothing to do with the shape or condition of the hose. All the hose does is provide a path for the water to flow through.

If I step on the hose, the flow is restricted. This would be likened to the mind of a person with very little Faith. The flow from Infinite Intelligence is obstructed by the lack of Faith. That small, quiet, intelligent voice that hands over new concepts, ideas, and solutions is not heard.

However, if I were to take my foot off the hose and allow the water to run freely, it is like the mind of a person who has a Faith-filled mind. The flow of ideas, plans, and thoughts from Infinite Intelligence flow freely and effortlessly.

Notice that in both cases, nothing changed with the state of Infinite Intelligence, i.e. the source of the water flow. It was always there. What determined how much water flowed was the amount of Faith present in the mind.

Some people then ask, "How then do I produce Faith?" Well, Faith works like this: The human mind cannot tell the difference between the "truth" and a "lie". Because of this, you can program your mind to have Faith and believe in anything you desire to believe in. The more you hear something, the more your mind begins to regard it as fact, whether it is true or not. And all of your actions are based on what you believe. *Human beings do not believe what we see, we see what we believe.* So the key is to literally... just believe. If you do that, Faith will come too.

In addition, there is a direct correlation between the time at which we receive an idea, the time it takes to act on that idea, and whether our actions are effective. The truth is, the longer you wait to take action after receiving an idea or plan, the less effective your actions will be when, or if, you do ever take action. The time it takes for you to move forward on an idea is a direct reflection of your Faith.

Matthew 9:29 states, "*According to your Faith,* let it be to you." The only reason people don't take action is because of indecision. Any indecision that you have will delay your action, and as we already know, indecision is the seedling of fear.

That is why the time it takes for you to act on an idea is so important. It is a measure of your Faith. The more time between an idea and action equates to less Faith. The less time between an idea and action equates to more Faith. "*According to your Faith,* let it be to you."

According to *Think and Grow Rich* by Napoleon Hill, Faith is a state of mind that can be developed by the persistent application of "autosuggestion". Autosuggestion is self-suggestion, or deliberately putting thoughts of your own choosing into your mind. Although these thoughts and suggestions must be repeated many times before they are accepted as Faith, once they are accepted, your actions will be automatic since you cannot act in a way that is in direct violation of what you believe.

All of us believe that we will not float if we step off the ledge of a tall building. Therefore (assuming we want to live), we will not purposely step off the ledge of a tall building. In the same sense, we will not jump into a deep river if we believe that we cannot swim. Our beliefs are formed over years and years from experience and self-suggestions. Any time emotions are added to our self-suggestions, they tend to "color" those beliefs and they will be transformed into Faith more quickly. Whether these emotions are positive or negative does not matter. The mind will accept a thought colored with a positive emotion just as readily as it will accept a thought colored with a negative emotion. Remember, the mind does not discriminate. It does not know the difference between "positive" and "negative". It simply accepts what you feed it as being true. So feed it positive thoughts, beliefs, and intentions!

Remember this: if you are not where you desire to be in your life, you have the divine right to change the course of that life by the use of autosuggestion. You have the absolute right to plant in your mind "little white lies" that you are successful, rich, ambitious, handsome, beautiful, courageous, or whatever you want to be—even if other people say you are not, or do not believe it. Autosuggestion is the only way to change your present situation. You must input different positive thoughts into your mind until those new thoughts take root and your actions eventually change.

This all carries over into your everyday life, because your everyday life is a product of what you have Faith in. If your everyday life is not what you desire it to be, you must change what you have Faith in. Only by changing your Faith, will your actions then change too, and ideally, for the better.

CHAPTER KEYS

- Ideas from Infinite Intelligence "flow" into your mind in much the same way that electric current flows in a circuit. Faith is what allows this flow to take place.

- Faith is the opposite of fear, and stronger than fear; your Faith must be stronger than your fear.

The Life Circuit™

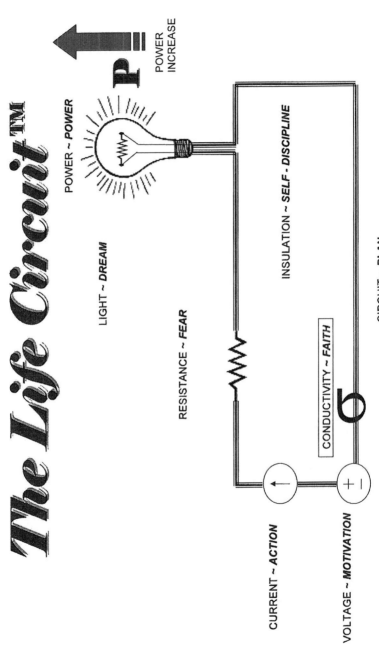

POWER ~ *POWER*

POWER INCREASE

P

LIGHT ~ *DREAM*

INSULATION ~ *SELF - DISCIPLINE*

RESISTANCE ~ *FEAR*

CONDUCTIVITY ~ *FAITH*

CIRCUIT ~ *PLAN*

CURRENT ~ *ACTION*

VOLTAGE ~ *MOTIVATION*

CHAPTER 9

SHORT CIRCUIT - Complacency/ Conformity/ Comfort Zone

ANALOGY: *A Short is to the Electric Circuit as Complacency is to your Life Circuit*

INTERLATION: "Short" ~ "Complacency"

Smooth seas do not make skillful sailors

- African Proverb

Triumphs without difficulties are empty. Indeed, it is defaulters that make the triumph. It is no feat to travel the smooth road.

- Author Unknown

A short circuit is essentially a connection made from one end of the voltage source directly to the other end. When a circuit is shorted, there is no resistance to impede the electric current flow. The electric current is flowing through the conductor with virtually no resistance. In other words, there is no light bulb connected in the electric circuit for the electric current to flow through. The electric current is actually being pushed around the electric circuit with no light to even turn on. Seems kind of strange, huh?

Well, because of the great amount of unimpeded electric current flow through the conductor, the conductor begins to burn. We must keep in mind the purpose of moving electric current is not just to get it to flow. We want electric current to flow towards a destination point, and in the example we have been using in this book, the destination point is the light bulb.

Remember that we use the act of turning on the light to represent the attainment of a dream in life. In a circuit, electric current (your actions) moves toward the light bulb (your dream) until the light turns on (your dream is attained). But if there is no dream in your life, it is just as if there is no light bulb to turn on in the electric circuit. If there is no light bulb to be turned on in the electric circuit, then what is the point of the electric current (your actions)?

There is no point, because the electric current must not only be moving, but moving *through some resistance* in order to create the power that turns on the light bulb. Electric current moving through a conductor with no destination point is futile, and it is what we call a "short circuit" condition.

The vast majority of people want, or think they want to get to a place in their life where they have no worries, no challenges, and no obstacles. And while it is true that life is not meant to be one continuous struggle with little or no reward, obstacles are a natural part of life. They must be dealt with accordingly, because they are the exercise that allows us to improve continuously.

For every setback there is a *seed* of equal or greater benefit.

Anytime we are experiencing a hardship or obstacle in life, there is a certain amount of fear involved. What we have to realize is that we are being molded and refined by the difficult situation we are facing. And the degree to which we grow—because of the obstacle—is a result of our ability to act despite our fear.

While they may not be pleasant at the moment, our obstacles and fears are necessities because without them we have nothing to go through. We have nothing to make us stronger. So, while on the outside it may appear to be nice to have no obstacles and fears, we really do ourselves a disservice by not having some sort of fear to challenge us; some sort of fear to work through and conquer.

Let us think for a second about anything that grows strong. How does a bodybuilder make his muscles big and strong? He lifts weights—big heavy weights. He cannot sit in front of the television doing repetitions of lifting the remote control and expect to be a world-class bodybuilder. Muscles won't grow by lifting small objects that we come into contact with every day simply because there is no difficulty. If the muscle could talk, it would probably say, "Why should I get bigger and stronger if I don't have to? This remote control is so light that it's easy to lift it up." It has a valid point. In order to trigger the muscle to grow, there must be a reason to grow. And if you are making your muscle lift a heavy weight over and over, your muscle begins to think, "Hey, this is hard. I need to get bigger and stronger because doing so will make this load easier to lift." Therefore the muscle grows in response to the work it must adapt to. If there is no work to be done, there is nothing to adapt to. The muscle will remain weak and small.

If you have ever seen an athlete who has broken some part of his leg and has been in a cast for any appreciable time, you will also see this principle in effect. The leg that is not in the cast keeps the muscle because it is still active and working. The muscle thinks, "I have work to do so I need to be big and strong." The muscle in the cast is inactive and not doing anything. It is on "Easy Street" for six weeks. It gets complacent and loses strength, thinking, "Hey, I'm not being asked to do anything, so I'm just gonna chill!"

In conclusion, always be on the lookout for challenges and tests to see what you are "made of". This is the only way for you to grow. And remember, if it doesn't kill you, it can only make you stronger.

CHAPTER KEYS

- A short circuit condition in a circuit occurs when there is no resistance for the electric current to travel through. Therefore, no power can be generated.

- In life, a short circuit condition is complacency. Because you have no resistance (everything is easy for you), you do not have the opportunity to create power in your life.

The Life Circuit™

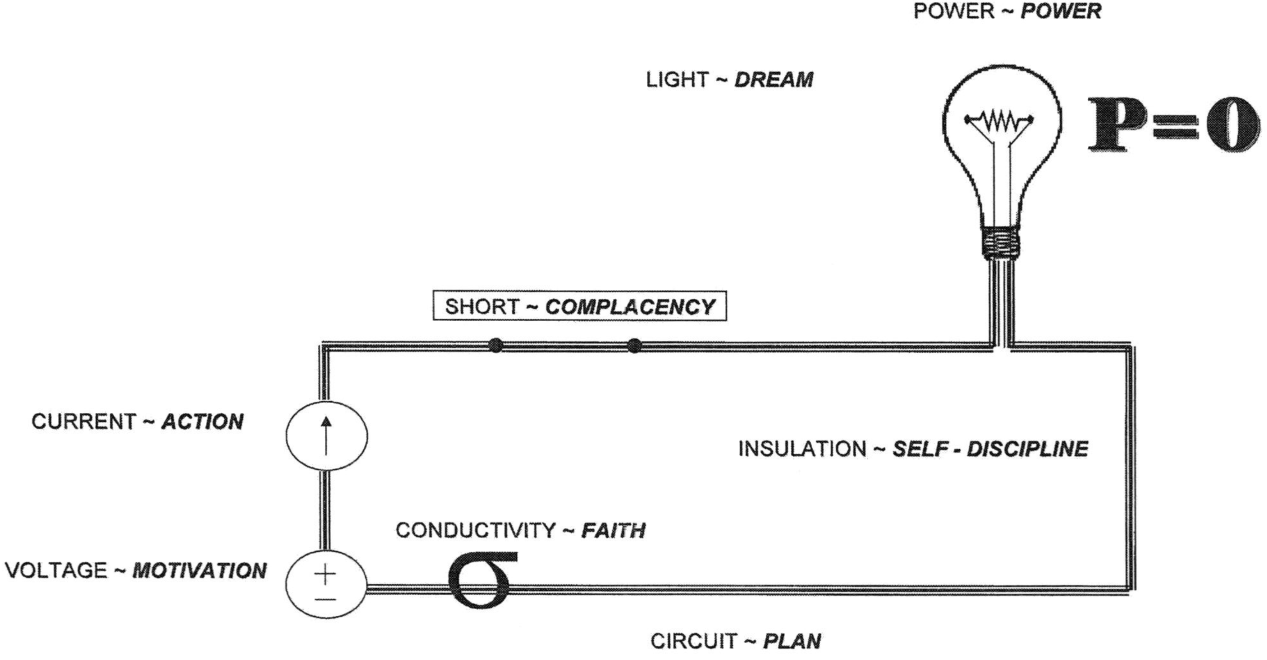

POWER ~ **POWER**

$$P = 0$$

LIGHT ~ **DREAM**

INSULATION ~ **SELF-DISCIPLINE**

SHORT ~ **COMPLACENCY**

CONDUCTIVITY ~ **FAITH**

CIRCUIT ~ **PLAN**

CURRENT ~ **ACTION**

VOLTAGE ~ **MOTIVATION**

CHAPTER 10

OPEN CIRCUIT - Paralyzed by Fear

ANALOGY: *An Open Circuit is to the Electric Circuit as Fear Paralysis is to Your Life Circuit*

INTERLATION: "Open Circuit" ~ "Paralyzed by Fear"

Life's challenges are not supposed to paralyze you, they're supposed to help you discover who you are.

- Bernice Johnson Reagon

Confidence is not acting without self-consciousness. Confidence is acting despite your self-consciousness. Similarly, courage is not acting without fear. Courage is acting despite that fear.

- Monroe Mann

In an open circuit condition, the electric current does not flow because there is a break in the path of the electric circuit. The circuit is open, and an open circuit has *infinite* resistance. This is bad, because infinite resistance is a fear that—for one reason or another—you cannot act through.

In Chapter Two, we learned how electric current is analogous to a person in action moving towards a pre-defined goal. Generally speaking, fear—more than any other single cause—is what prevents people from taking that action. Fear is the killer of action. But it doesn't have to be.

Let's take a look at how fear is born. Once we understand it, we can then act despite that fear. Otherwise our fear will grow until we become paralyzed. Remember, having fear is okay. But what is not okay is allowing fear to prevent you from taking action towards your pre-determined goal.

The Progression of Fear

Stage 1 - Indecision

The seed of fear, indecision, is planted in the mind. Many times this seed is planted subtly or without notice. It may land in the mind by way of some off-hand comment by a "friend" or relative. Or it may be planted there intentionally by enemies who want nothing more than to see you fail. In any case, this seed raises doubt in your mind. This is the crucial stage at which the seed of fear must be destroyed. We cannot *totally* control our environment; there will always be little seeds of indecision floating around, looking to take root in your mind. And some of them will land. Your job is to recognize this seed by being ever alert of how quickly you make your decisions. You can gauge the amount of fear seedlings trying to take root in your mind by analyzing how long it takes you to render decisions when necessary. Making quick and firm decisions indicates that seeds of fear are being eliminated from your mind. All fears are grown out of the inability to make a decision of some sort. Indecision leads to inaction. Decision leads to action.

Stage 2 - Doubt

Some time has now passed since the initial bout of indecision. However, a decision still has not yet been made, and as a result, no action has been taken either. Doubt is beginning to take over the mind, and good ideas are now being second-guessed. The longer you wait to make a decision, the stronger the doubt becomes in your mind. Also, you begin to rationalize why it takes so long to make a decision by trying to make a perfect decision or wait until the "time is right". However, at this stage, what you need to understand is that it doesn't have to be the best decision—in almost all situations you will have the opportunity to go back and revise it later. Therefore, making a decision in the first place is truly what is most important. It may be helpful to incorporate advice from a mentor or counsel at this time if you have not done so. Here, time is your enemy because it is growing the seedling of indecision into the root of fear.

Stage 3 - Fear

Fear takes root in the mind, where it begins to cripple your capacity to think and make rational decisions. At this point, you are in danger of not making any decision at all. Your Faith has been choked by your overpowering fear. Many times, you may be unaware that you are afraid of something. But you can always verify your fear by simply thinking about the things you know you need to do, but haven't yet done. Those are the things you are afraid of. Behind all inaction is some type of fear.

I have a belief that the most admirable quality a man can possess is the ability to act in the face of fear. If you discover that you have the habit of becoming overtaken by fear, here is a suggestion. Repeat the phrase "READY, FIRE, AIM" twenty times in the morning when you wake up, and twenty times at night before you go to sleep. Notice this is not "READY, *AIM*, FIRE."

READY - The best time to do something is always *now*. In other words, you are always ready, whether you think you are or not.

FIRE - Go ahead, and start it! Just do it!

AIM - You can adjust later. Think seriously for a minute. Most of the things you do, whether it be a report, presentation, project negotiation, whatever - you usually always have an opportunity to make revisions or in some way make the best of the decision you have made. Very rarely is the act of making the "perfect" decision more important that making a "timely" decision. You might want to re-read that last sentence.

Take note: the decisions you have to make in life tend to repeat over and over—disguised as different circumstances—until you learn the intended lesson. You will have many opportunities in life to get the decision "right", even though at the time it may seem like a one-shot deal. In other words: *Make your bold decision and go!* You will learn more using this method than wasting time being indecisive and inactive.

The problem with the alternative, "READY, AIM, FIRE," is that you usually spend too much time AIMING. There is nothing wrong with planning and analyzing a situation—planning is very important. But if you take too much time, indecision creeps in and you run the risk of becoming paralyzed by fear. All the best laid plans in the world are of no use if you cannot act on them because of fear. Successful achievers plan and organize, but they also make their decisions quickly. I can't say it often enough: indecision is the seedling of fear. Therefore making quick decisions prevents indecision from creeping in.

Please, do not get into the habit of trying to get everything perfect on the first try. Many perfectionists are plagued by inaction because they cannot seem to get going unless conditions are perfect. But if you wait until conditions are perfect before you begin, you will never get anywhere. You can't wait until all of the lights on the street are green to decide to move your car forward; if you did, you would never get anywhere.

Breaking down to the fundamentals, fear is the only thing that stands between you and your dreams. Got it? Truly, what would you do *today* if you knew you couldn't fail? Your mindset would be, "I have nothing to lose by going all out and giving my best effort!" However, most people's minds are so busy focusing on what will happen if they don't hit their goal that they don't even start—when they should be focused on how great their life will be once they *do* hit their goal. There really is nothing to lose. If you try and fail, you are still in a better position than if you hadn't tried at all because you now know what doesn't work. And discovering what doesn't work is just as much a part of success as discovering what does work. All truly successful people have *failed* their way to success.

Moving back to the world of electricity, if we were to create an open in an electric circuit, no electric current would flow. The electric current would not flow through the air because air is an insulator. There is no way for the light to turn on because there is an open in the electric circuit. The electric current is no longer flowing.

Now let's think about our own personal lives for a second. We are like electrons, and when we move, we are like electric current. In the case of the electric circuit, we looked at the open circuit being the reason the electric current stopped flowing. In our personal lives, why do *we* stop moving and taking action? We can come up with a thousand different "reasons" (excuses) why people stop taking action, but they all boil down to fear in one shape or another. Fear is why we stop taking action. We become afraid and we allow that fear to prevent us from moving—in the exact same way as an open breaks the path of electric current flow and prevents the electric current from flowing toward the light bulb.

Think back to any situation in your life in which you were afraid. Chances are, the fear grew in proportion to the amount of time you procrastinated before taking action. And chances are, you began to feel less afraid as soon as you began to take action. The more action you took, the less and less you were afraid.

This is the paradoxical aspect of fear, and why it is so deadly. Fear makes you want to *not* take action. However, the thing that cures you of your fear is—strangely enough—taking action.

CHAPTER KEYS

- Electric Current cannot travel if there is an opening in the circuit. In the same manner you cannot take action towards your goals and dreams if you are paralyzed by fear.

- When confronted with fears, the best choice is to always act in face of your fears. Otherwise, indecision will turn to doubt, which will then become a permanent fear in your mind.

The Life Circuit™

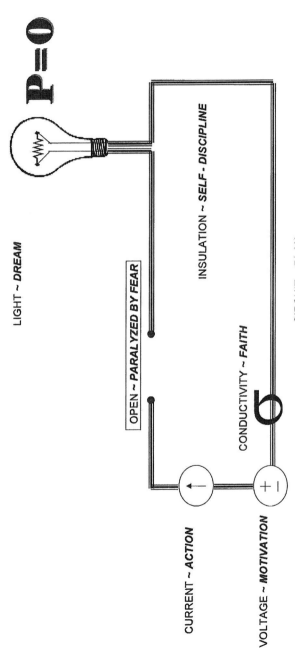

POWER ~ **POWER**

P=0

LIGHT ~ **DREAM**

INSULATION ~ **SELF - DISCIPLINE**

OPEN ~ **PARALYZED BY FEAR**

CONDUCTIVITY ~ **FAITH**

CIRCUIT ~ **PLAN**

CURRENT ~ **ACTION**

VOLTAGE ~ **MOTIVATION**

CHAPTER 11

CAPACITANCE - Imagination

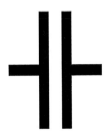

ANALOGY: *Capacitance is to the Electric Circuit as Imagination is to Your Life Circuit.*

INTERLATION: "Capacitance" ~ "Imagination"

I am imagination. I can see what the eyes cannot see. I can hear what the ears cannot hear. I can feel what the heart cannot feel.
- Peter Nivio Zarlenga

At any given point in time, we lie in the midst of tremendous opportunities for achievement, if only we have imagination enough to grasp hold of them. Without the capacity of our imagination, those opportunities are simply wasted; they mean nothing to us.
- Kolie E. Crutcher III

Electrical capacitance is the ability of an object or surface to store an electrical charge. It is simply a measure of the electrical storage *capability* of the object.

Capacitance is analogous to the ability of a jar to hold a fluid. Just as some jars have more capacity than others, some objects have more charge-holding ability than others. Thinking along those same lines, some people have more ability to achieve than others. A jar's ability to hold a fluid is limited by its physical size. An electrical capacitor's ability to hold charge is limited by its capacitance (which is related in part to its physical size).

However, a human being's ability to achieve is limited only by his or her *imagination*. The reason there is no limit to how much we can achieve is that there is no limit to our imagination.

You see, the mind of every human being is similar to a jar that can be varied in volume. A small imagination has room to hold only small achievements. But when you stretch your imagination you increase the volume of this jar, and you have the ability to hold more achievement. And when you stretch your imagination again, you further increase the volume of this jar. Now you have the ability to hold even more achievement. The wonderful thing about an imagination is that there is no limit on how large it can be stretched! Therefore you will never run out of room to achieve, unless you choose to stop stretching and growing your imagination.

Try this simple experiment: Take a large glass and fill it with water. Then take a smaller glass that is empty and set it next to the large glass filled with water. Begin to pour the water from the large glass into the small glass. Don't stop until the large glass is completely empty. Now look and see how much water is in the small glass. What you will see is that no matter how much water the large glass held, the small glass is limited as to how much of that water it can hold. The small glass has the ability to hold only a small amount of water. The remainder of the water wastes out onto the table. When we do not use our imagination to increase our capacity to achieve, we are like the small glass. We lose all that opportunity to achieve simply because we don't have the capacity to hold on to it when it comes to us. Expecting to achieve greatness while possessing a small imagination is like expecting to pour a gallon of water into an 8oz glass and not expecting to spill a drop!

You will never achieve greatness until you can first see greatness without your physical eyes—until you see greatness in your imagination.

Imagination leads to the Special Theory of Relativity

Perhaps the most famous mathematical equation ever formulated is $E=mc^2$, which is a surprising consequence of Einstein's Special Theory of Relativity. Even people who are laymen in the fields of science and mathematics have heard of this equation. But what does this equation really mean, and where did it come from?

As a child Albert Einstein asked himself, "What would it be like to travel alongside a wave of light?" When he got older, he discovered it would be very difficult to do so because of the extreme speed at which light travels—186,000 miles per second. So Einstein used his imagination to picture himself traveling at that speed, and that is how he discovered relativity. Stated simply, Einstein's Theory of Special Relativity says that distance and time depend on the observer; they are not absolute. Also time and space are perceived differently, depending on the observer. In addition, Einstein questioned what would happen to things as they traveled at such extreme speeds.

Years earlier, Dutch physicist Hendrik A. Lorenz found that electrons gained considerable mass when traveling at speeds near the speed of light. This was measured in a cathode ray tube, similar to the kind that produces pictures in traditional television sets. This puzzled Lorenz, because up until that point, most physicists were using the laws of Isaac Newton. One of Newton's laws stated that the mass of an object was a constant. However, the mass gain of the electrons moving near the speed of light violated this law, and Albert Einstein discovered why.

Einstein theorized that as a particle was forced to move quickly, it gained energy. Therefore if a particle was moving at an extremely fast speed—such as the electrons traveling near the speed of light—those electrons must gain a great deal of energy. He thus concluded that energy has mass. This shocked scientists around the world because energy was never thought to have mass.

Einstein then went one step further with his research and formulated the equation $E=mc^2$. The "E" stands for energy. The "m" stands for mass. And the "c" stands for the speed of light. Einstein's theory states that mass is simply "frozen" potential energy. The amount of energy within a few pounds of coal is vastly more than the amount of energy released by burning that same few pounds of coal. There would be roughly 25 billion kilowatt hours generated by *releasing* the stored energy as opposed to the small amount of energy generated by *burning* the coal. The problem was: how to release stored energy from matter in a manner that is safe and controlled? The answer? Nuclear power, which deals with methods to achieve this.

What's my point? My point is that creativity and imagination are perhaps the best tools for solving complex problems, and Albert Einstein used these tools on a regular basis. Albert Einstein reorganized the knowledge that he gained from school and books, and using his imagination, connected concepts in his head that could not be realized in the physical world, and created pictures in his mind's eye that human beings just couldn't see with the naked eye. Creativity and imagination were very crucial to Albert Einstein and his success. He thought about mass and energy in a new, creative way, which contributed greatly to him becoming arguably the most famous scientist of this century.

Perhaps the most famous quote by Albert Einstein is "Imagination is more important than knowledge. Knowledge is limited. Imagination encircles the world." He probably came up with this quote while cutting class (which he did quite often). Instead of attending traditional classes, Einstein chose instead to spend his time in his room reading and stretching his imagination to the capacity that allowed him to become the "Einstein" we know today; to in fact become the standard by which—it is important to note—all future geniuses would be measured.

CHAPTER KEYS

- Your ability to achieve is limited only by your ability to imagine.

- You must use your imagination to create the image in your mind before it will become a physical reality.

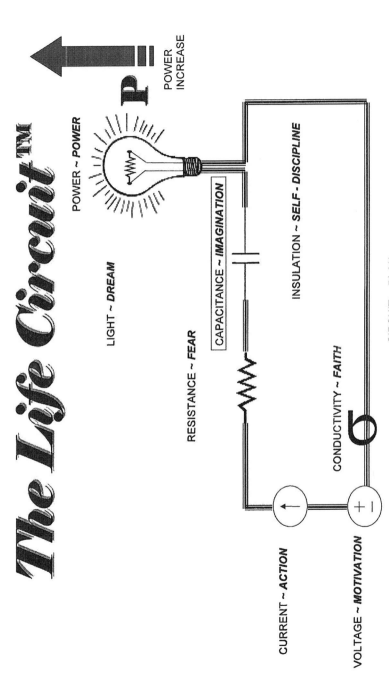

The Life Circuit™

POWER ~ **POWER**

P POWER INCREASE

LIGHT ~ **DREAM**

CAPACITANCE ~ **IMAGINATION**

INSULATION ~ **SELF - DISCIPLINE**

RESISTANCE ~ **FEAR**

CONDUCTIVITY ~ **FAITH**

CIRCUIT ~ **PLAN**

CURRENT ~ **ACTION**

VOLTAGE ~ **MOTIVATION**

CHAPTER 12

INDUCTANCE - Persuasion

ANALOGY: *Inductance is to the Electric Circuit as Persuasion is to Your Life Circuit.*

INTERLATION: "Inductance" ~ "Persuasion"

The power which electricity of tension possesses of causing an opposite electrical state in its vicinity has been expressed by the general term Induction.

- Michael Faraday

I would rather try to persuade a man to go along, because once I have persuaded him, he will stick. If I scare him, he will stay just as long as he is scared, and then he is gone.

- Dwight D. Eisenhower

Back in 1831, Michael Faraday made a very important discovery. He determined that by moving a wire within a magnetic field it was possible to induce electric current in the wire. Up to this point, we have only known of one method to cause electric current to move through a conductor: the connection of a voltage source to the conductor as discussed in Chapter Three.

However, Faraday's discovery showed that electric current could be caused to flow in a conductor by using another method: **whenever a conductor is moved in a magnetic field, electric current flows in that conductor too.** This electric current is called an *induced* electric current. Without going into too much detail (because the study of magnetism is a science in itself), a magnetic field is produced whenever the atoms in a ferromagnetic material are aligned. And now your next question is naturally, "What is a *ferromagnetic* material?" Here is the short and sweet explanation.

Everyone knows the general concept of a magnet. Magnets "stick" to some metals. Iron, nickel, and alloys containing either or both of these elements are known as *ferromagnetic* materials. When a magnet is brought near a piece of ferromagnetic material, the atoms in the material become lined up, so that the material is temporarily magnetized. This produces a magnetic force between the atoms of the ferromagnetic substance and those in the magnet.

If a magnet is brought near another magnet, the force is even stronger. Not only is it stronger, but it can actually be either repulsive *or* attractive, depending on the way the magnets are turned. And this force gets stronger as the magnets are brought nearer to each other.

Ok, there you go. That is enough of a background in the study of magnetism to understand what a magnetic field is and how a magnetic field is produced. Now, back to the original reason we were interested in magnetic fields—the fact that moving a conductor within a magnetic field induces an electric current to flow in that conductor.

Now that we have a basic understanding of how induction current is produced, we can make the connection between induction in the field of electricity and persuasion in the field of life. The connection is this: *In order to become powerful, you must be able to induce action in others.*

According to *www.Answers.com*, the word "induce" actually means, "to lead or move, as to a course of action, by influence or persuasion."

Bottom line is this: You, as the up-and-coming powerful achiever, must be able to *induce* others to take action on your behalf to achieve your dreams. If the *only* actions taken toward your goals and dreams are those actions taken by you yourself, you won't get much done. Understanding the fact that you are only one person, and as such only have 24 hours a day in which to work (if you never took a break), it becomes clear that the achievement of major scale dreams involves the cooperation and help of others. That is why you must be an inductive (persuasive) person.

All great and noteworthy achievements were born in the mind of the dreamer and came into reality because the dreamer was able to induce in others the *collaborative* **actions** necessary to generate power. Remember from Chapter Six that there is no power without action. And the more action you take toward your goal, the more Power you generate toward the attainment of that goal.

J. Paul Getty, the US oil industrialist, once stated that he'd rather have 1% of 100 men's efforts than 100% of 1 man's effort. Getty was right. Think about how much more power you can generate by inducing others to take actions on your behalf as opposed to trying to do everything by yourself!

Let me make myself perfectly clear when I talk about the need to induce and persuade other people to take action on your behalf. I am **not** saying that you be lazy and sit back and just try to tell others what to do. That won't work. You must know how to work in teams, and as a leader. In order to be a persuasive person, you must also be highly self-motivated. No person is ever going to willingly do something for any leader when that 'leader' would not consider doing that very same job himself. It's almost as though your self-motivation needs to become so hot, that like a fire it spreads to others. These others will then ignite too, and want to act, because of the spark that you have drawn out from within them.

And let's not forget the fact that presumably, you are striving to achieve some worthy ideal of great proportions. As such, there will undoubtedly be many different parts making up the complex whole. You, as a single person, have neither the time nor the need to become an expert at all. Many people from different disciplines possessing different expertise must be coordinated to achieve noteworthy success. Again, you must work in teams. If ever in doubt, remember the words of J. Paul Getty. "I would rather have 1% of 100 men's efforts than 100% of 1 man's effort."

The Difference between a Leader and a Manager

In essence, a leader is a master of the art of persuasion. People follow leaders and do things for leaders because they *want* to do them. That leader has inspired something within that other person, and now the person is acting because he or she has the desire to do so. A leader who creates these types of followers never has to worry whether work will continue to be done in his absence.

On the other hand, a manager gets people to work out of a sense that they *have to* work. The problem with managing people as opposed to leading people is that the manager must always be around to make sure everyone is working as they should. Once the manager stops supervising, the people stop working.

An Example of Induction

Are you inducing electric current in others or is your electric current being induced? It is either one or the other. This is a good question to ask yourself periodically. By answering honestly, you will have a very good idea of whether you are being Proactive or Reactive. Remember from Chapter Two that Proactive people get rich. Reactive people get what is left over after the Proactive people are done.

On that note, there may be some slight confusion regarding the difference between voltage and induction. Some may say, "Both voltage and induction cause electric current." The following may be a simple way to remember the difference in the two:

Your Voltage (Motivation) causes *your* Electric Current (Actions).
Your Induction (Persuasion) causes Electric Current (Actions) in *others*.

Stated another way:

YOUR MOTIVATION DETERMINES WHAT *YOU* DO!
YOUR PERSUASION DETERMINES WHAT *OTHERS* DO **FOR** YOU!

CHAPTER KEYS

- To *induce* is to *persuade*. You must be an inductive person to achieve noteworthy success because you, by yourself, are limited to 24 hours per day to be productive.

- When inducing others, you must understand that people do things for *their* reasons, not yours. It is up to you to find their reasons, so that that you may successfully induce them.

The Life Circuit™

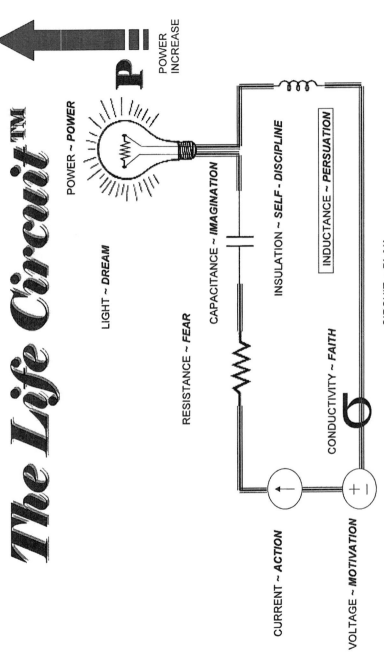

POWER ~ **POWER**

LIGHT ~ **DREAM**

POWER INCREASE

RESISTANCE ~ **FEAR**

CAPACITANCE ~ **IMAGINATION**

INSULATION ~ **SELF - DISCIPLINE**

INDUCTANCE ~ **PERSUATION**

CONDUCTIVITY ~ **FAITH**

CIRCUIT ~ **PLAN**

CURRENT ~ **ACTION**

VOLTAGE ~ **MOTIVATION**

CHAPTER 13

FREQUENCY – Attitude!

ANALOGY: *Frequency is to the Electric Circuit as Attitude is to your Life Circuit*

INTERLATION: "Frequency" ~ "Attitude"

A great attitude does much more than turn on the lights in our worlds; it seems to magically connect us to all sorts of serendipitous opportunities that were somehow absent before the change.

- Earl Nightingale

Have you ever thought about the technological principles that allow you to actually listen to your favorite radio station? Technology is so advanced today that we give little thought to the details of how we are able to enjoy the things that are available at the touch of a button.

In short, the theory of listening to your favorite radio station involves a sending station, a receiving station, and a sine-wave signal. The sending station takes a message (maybe a human voice or music) and encodes that message onto a sine-wave signal. That sine-wave signal is then *stepped-up* to a very high frequency so that it can be transmitted over long distances. This stepped-up sine-wave signal is not audible to the human ear, but necessary for the encoded message to be transmitted. A receiving station that is "tuned-in" (to the frequency that the sending station is sending out) can "pick-up" that signal. Once the signal is picked up at the receiving station, it is stepped back down to a lower frequency where the original message is once again able to be heard as a human voice or music.

When you "tune" your radio dial, you are actually selecting which frequency you want to "pick-up". Each different radio signal uses a different sine-wave frequency. That is how they are kept separate from each other.

When you "tune" your radio dial to 97.1, you are telling your radio receiver "I want to pick-up the radio station that sends out its messages on a sine-wave frequency of 97.1 MHz." The radio receiver then picks up this signal. But since 97.1 MHz is a frequency much too high to be detected by the human ear, the receiver must also step this signal down to the original message frequency so that you can hear it.

Now that you understand the basics, frequency is measured in something called hertz, or cycles per second. In other words, a signal that has a frequency of 1 hertz goes through one cycle every second. Without getting overly technical, typical signals may look like this:

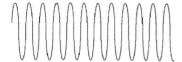

High frequency – ENTHUSIASTIC ATTITUDE

Low frequency – PESSIMISTIC ATTITUDE

Figure 13.1 – Two signals with different frequencies

The first signal has a high frequency, while the second signal has a low frequency. If you could see enthusiasm in your life, it would look very similar. You will achieve very little in your normal 'frequency' state. You must first step yourself up by becoming enthusiastic about your goal and mission so that you can travel great distances and achieve great things. Show me a person who has no enthusiasm about his or her work and I will show you a person who is bound to fail. There is simply no substitute for enthusiasm. It is a force that sweeps obstacles to the side.

Some people may say, "I just don't *feel* enthusiastic." These are people who do not understand the principle of Inner First, Outer Second. This principle says that how you *feel* can be a function of how you *act*. Most of us know that it is relatively easy to feel a certain way based on the things that happen to us. We win a vacation and we are happy. Someone close to us dies and we are sad. A friend betrays us and we are hurt. And we tend to *act how we feel* about these situations until another situation comes along that changes how we feel. Then we act accordingly to how we feel about the new situation. We are merely *along for the ride*, hoping that circumstances are in our favor.

The problem with this, however, is that we are allowing circumstances to control how we feel. If life were always nice and great, this would not be a big deal. But we all know that life deals everyone blows and knockdowns, especially those who are striving to achieve. Therefore, when the circumstances change for the worst we are in trouble because we have given control of our attitude over to the circumstances.

Mental power, then, is the ability to create your own enthusiasm, *regardless of the outside circumstances.* When you harness this principle, you have taken the power away from the circumstances and regained control for yourself. You are no longer dependent upon good circumstances in order to feel good, and you do not become overtaken with bad feelings and depression when circumstances are unfavorable. This is true enthusiasm because it is created from within, and totally within your control, and not influenced by outside events.

Now, doesn't it seem very worthwhile to develop that enthusiasm through your own mental power, knowing that enthusiasm is a necessity for achievement in any field? Wouldn't it be nice, and make more sense, if you could call it up whenever needed, as opposed to waiting for favorable circumstances which may or may not come?

How to Become "Enthusiastic!"

The secret to mental power and becoming enthusiastic is realizing that *you have the ability to change how you feel by changing how you act.* Your actions affect your feelings!

These actions may feel unnatural at first, and they are certainly not automatic because it is not how you genuinely feel. And that is precisely why you must impose them on yourself until you actually begin to feel different!

And once you do feel differently you *automatically* will act differently as well. This is the very opposite of the majority of people's view on enthusiasm, because they allow their feelings to change their actions.

It is an automatic response. That is why you must always be aware of how you are feeling and keep your level of enthusiasm high. If your enthusiasm begins to drop, you must first realize that this is the natural tendency, and then, you must impose your mental power and *act with an enthusiastic attitude* until you actually *feel enthusiastic!* Then, once you really feel enthusiastic again, you will automatically be enthusiastic again. You see, if you act enthusiastically in the meantime, you will also eventually start to feel enthusiastic again as well.

Monroe Mann has a mantra that only two things matter for success. Those two things are 1) ATTITUDE and 2) BUSINESS SENSE. Talent is assumed, and therefore relatively unimportant. If you read his books *The Theatrical Juggernaut, Battle Cries for the Underdog, To Benning & Back,* or his upcoming new one, *The Artist's MBA,* you will find this theme running strong throughout them all. In *The Theatrical Juggernaut,* Monroe is talking about attitude and business sense particularly in the field of acting and the arts. However, this is true in *any* field.

In any field you choose to endeavor, the only two things that are important to success are your attitude and your business sense. If you have both of those qualities in great supply, and continually work to improve both, you will become successful—whether in show business, or in life. And as Monroe makes clear, your *attitude* is the most important quality of all: No Rules. No Excuses. No Regrets, as he says.

It is so important for you to understand this. Your attitude is completely determined by only one person—YOU. Your attitude can either be positive (enthusiastic high frequency) or negative (pessimistic low frequency). That attitude is the ONLY thing you have complete control over in this world. This is not by mistake. There is a reason that our Creator gave us complete control over only one thing on this earth (our attitude). That is, coincidentally, the only thing that matters. An enthusiastic high frequency attitude is how we combat the negative things that happen to us in our lives (since we cannot always control them directly).

The relevance and importance of this awesome power to adjust and control our own attitude is demonstrated by its importance to two of the most powerful and influential achievers the world has ever seen: Thomas A. Edison and John D. Rockefeller, Jr. Both of these gentlemen were also well known to live by similar attitude-focused mantras.

Let's first talk about Edison. For those of you who may not be aware, Thomas Edison was arguably one of the most productive inventors and businessmen this country has ever seen. He earned 1,093 United States patents and has been credited with improving the first reliable light bulb, the motion picture machine, the phonograph, as well as many other notable inventions. Edison's influence was far reaching. As a matter of fact, Henry Ford's most noticeable advances did not come until he formed a business and personal relationship with none other than Thomas Edison. (On a side note, one of Henry Ford's great quotes was, "Whether you think you can, or you can't, you're right.")

Edison attributed his great success to "TIISG", which was an acronym for "Turn It Into Something Good". Edison truly believed that every setback and every temporary failure could be turned into a positive situation of equal or greater benefit. Because of this positive attitude (enthusiastic high frequency attitude), Edison was unstoppable in his business ventures. Whereas others may look at hard times, setbacks, and failures as signals to give up, he saw them as opportunities to TIISG—to turn it into something good. Edison realized that in every situation, he had a choice in deciding how he reacted to what happened to him. He probably heard the fable about the goat in the ditch that I shared in the beginning of this book!

One of the greatest examples of this occurred in 1914, when the entirety of Edison's West Orange, NJ laboratory was destroyed by a fire. Although the damaged exceeded $2 million, the building was insured for only $238,000. Edison's 24-year-old son Charles was devastated. But Edison quickly analyzed the situation and stated, "There is great value in disaster. All our mistakes are burned up. Thank God we can start anew."

Three weeks later, Edison broke through with the world's first working phonograph.

Just like the goat, Edison realized that the circumstances of life that appeared to be piling on top of him and weighing him down could be used as stepping stones to bring him to a higher level—if he simply chose to look at them in that way. The key, as you can see, is the attitude of the person experiencing the circumstance, and not the circumstance itself. The circumstance itself is unchanging; the way you look at the circumstance is what you have full control over.

The things that make life seem bad when they are on top of you are the same things that make life seem great when *you* are on top of *them*. Edison was fully aware of this, and he took it upon himself to understand how to take every setback or temporary failure and TIISG: TURN IT INTO SOMETHING GOOD!

Attitude and business sense are the only two things that count!

Let's move our attention now to John D. Rockefeller for a moment. John D. Rockefeller created Standard Oil which was at the time the largest business empire on earth. In 1910, Rockefeller's personal wealth accounted for 2.5% of the United States economy. John D. Rockefeller was once quoted as saying that his goal was to turn every disaster into an opportunity. Again, this is the attitude of enthusiastic high frequency, and—you might note—very similar to Edison's TIISG principle.

To put the wealth of John D. Rockefeller into prospective, if you were to translate his money into today's dollars, Rockefeller would be worth an astounding $320 *billion*. That's *billion* with a "B". This figure puts Rockefeller at the top of the list as the richest man who ever lived in modern history! To put this into even more perspective, Microsoft mogul Bill Gates is *only* worth $59 billion.

When William Shakespeare said, "There is nothing good or bad, but thinking makes it so," he must have heard the story about the goat in the ditch as well. Is the dirt that is being piled on top of you a good thing or a bad thing? Neither. The circumstance itself is neutral. Your thinking—your mental response—is what determines the final outcome.

So again, whether your field is acting, inventing, oil, whatever - the only two things that are important are 1) Your Attitude and 2) Your Business Sense. Your attitude is the only thing that is entirely up to you, and an attitude of enthusiastic high-frequency energy will lead you to the opportunities that will ultimately sharpen your business sense.

What are you "Tuned In" to?

Let's return for a moment to the principle of listening to a radio station. Each human mind can be likened to both a "sending" and "receiving" station. And the frequency of each mind is the attitude. This is the reason people naturally seek out and are attracted to other people who are most like themselves. Whether we realize it or not, our minds are at any given time at some point along the "attitude scale."

You can think of one end being pessimistic low frequency and the other end being enthusiastic high frequency. We are all constantly sending out our attitude on its frequency. At the same time, we are constantly walking around with our antennas up in an attempt to pick up signals sent out by other minds at the same frequency.

Contrary to popular belief, opposites do not attract. People who wake up at 5am do not hang out with people who wake up at noon. People who are dedicated to going to the gym and eating healthy do not hang out with chocolate cake eating, soda guzzling couch potatoes. Even in the Bible, there are warnings about being "unevenly yoked". You will never find a person who has an enthusiastic high frequency attitude voluntarily hanging out with a person with a pessimistic low frequency attitude. They will not "hear" each other. They will become irritated. It would be like someone wanting to listen to 97.1, and not realizing that the radio receiver is set at 90.6. If the sending station and the receiving station are set to two different frequencies, they will not hear what they desire to hear.

Think about phrases such as "Birds of a feather flock together", "It takes one to know one", "Misery loves company", etc... These are all true, and they are different ways of saying the same thing: People who hang out and do well together are sending and receiving on the same mental frequencies. Right now, given that you've made it this far through the book, I can assume that my mental broadcasting station is sending out at the same/similar frequency that your mental receiving station is set to receive, and vice versa. Therefore, we "hear" each other. We are on the same page. We understand each other. We're essentially agreeing with each other, saying, "I want to listen to what you are sending out and you want to listen to what I am sending out."

I know you've probably heard this before, but if you want to know what type of person you are, simply look at your environment. Look at the things around you. Look at the people you hang out with. Those people are around you because you are all set to the same or similar mental attitude frequency. Since the mental attitudes are set to the same frequency, you hear each other. You are attracted to each other precisely *because* you are all listening to the same clear station.

Have you ever noticed that when you want to make a dramatic change in your life (remember a change in your physical life cannot happen without that very change first happening in your mind), you start to lose friends? My good man Monroe talks about this constantly using the analogy of your life as a train that continually moves faster with accomplishment. A lot of the people who started off with you on the train end up jumping off as you start moving faster and faster and faster (i.e. as their fear starts to grow, they bail)—and there's nothing you can do about it *except* be looking for the new (and 'high frequency') people who are jumping on at the higher speeds!

Here's an example: Let's say that for years you have been drinking, partying and living life "on the edge". Then, you want to change your lifestyle. You want to sober up, become organized and responsible. Inevitably, the first reaction from your old friends will be, "Oh, he thinks he's too good for us now!" Okay, maybe you are, but the real issue is that you have changed your mental frequency and are now "out of tune" with your old friends who are set at the same old frequency.

Here is another example of problems caused by different frequencies. If you go on a vacation to Europe and you bring your hair dryer from the United States, you may be in for a surprise when you jump out of the shower and attempt to dry your hair. Your hair dryer may not work. Why is that? In the United States, the electricity sent to the wall outlet is at a frequency of 60 Hz (short for Hertz. Remember that?) and therefore, the appliances we use in the United States are designed so that they operate at a frequency of 60 Hz. They match up. However, in Europe, the frequency of the electricity in the wall outlet is not 60 Hz; it is 50 Hz. So when you attempt to plug your American hair dryer (designed for operation at 60 Hz) into a European wall outlet (supplying electricity at 50 Hz), they do not match up. The hair dryer will not operate because the electricity required to operate it is at too low of a frequency. Make sense?

What you must come to understand is that *it's not personal!* Many people think that life is somehow out to get them. They think that just because the situation or circumstance happened to *them*, then it's personal.

Life is not out to "get" you. *Life* doesn't care whether you win or lose, succeed or fail. The only reason it appears that way is because most people have mediocre and negative habits, which inevitably lead to mediocre and negative results. Therefore, they feel like everything is against them. It is like constantly swimming upstream against the river current: your habits get you no where. Instead, try swimming to the river bank and walking instead, or finding a boat with an engine!

Consider this: the current is going where it is going to go. The same current that causes one to struggle and drown is the same current that helps another move swiftly and efficiently. Life is a constant current. When you find yourself struggling, and feeling as though you are on the verge of drowning, just remember: it's not personal. That same downstream current is moving someone on to the heights of prosperity, i.e. someone who is actually trying to go downstream!

It is your responsibility to learn how to use the current to move you swiftly forward instead of drowning you. Here's a case in point: there are quite a few professional swimmers these days who use current pools as a training device: it makes them stronger, and allows them to swim non-stop without having to turn around. The current pushes them back, and they swim forward to stay put. Yes, in this situation, the whole point is *not* to move.

Remember, the current doesn't know you and the current doesn't care about you! The current is going to flow as it flows, regardless. I'm going to repeat the quote by William Shakespeare again. "Nothing is good or bad, but thinking makes it so."

TIISG Correlation

A correlation to TIISG is that *when one door closes, another one opens.* The problem—as Helen Keller once pointed out—is that when the first door closes, many people are too busy being depressed and miserable due to their "misfortune" that they never notice the door that has opened. If you hang your head in defeat, how can you possibly see another door that has opened right in front of your eyes?!

If you can discipline yourself into TIISGing your so-called bad circumstances, i.e. turning them into something good, you will be utterly amazed at what you can accomplish. As a bonus, the world will be utterly amazed at what you accomplish as well. Furthermore, you will be training your mind to operate at the same attitude frequencies as Thomas Edison, John Rockefeller, Monroe Mann, and Kolie Crutcher. How exciting is that?!

To help you along in this process of TIISG, I have included a "TIISG (Turn It Into Something Good) Chart in Appendix III of this book. Tear it out. Make copies. Fold it up and carry it with you everywhere you go. This chart is to be *used*. Its purpose is not to sit in the back of this book and look pretty. That will do you no good.

<u>What do you see?</u>

Maybe you guys have seen this picture that has been circulating in the public domain, or a picture similar to this somewhere before. Either way, take a look, and tell me what you see.

Figure 13.2 – Ugly Old Witch or Beautiful Young Girl?

Is the picture an ugly old witch, or a beautiful young girl?

The answer is neither, or both, or that depends, or something else. My point here is that who's to say what this is a picture of? In essence, this is nothing more than a pattern of ink on paper. It is nothing at all until our minds *make it into something*. One person's mind will "see" an ugly old witch. Another person's mind will "see" a beautiful young girl. Another person may look at this picture for an hour and see neither. Your mental attitude is the determining factor.

In life, the circumstances we face are the pictures. They are completely neutral. Whether we choose to see those pictures as an ugly old witch or as a beautiful young girl is up to *us*. The frequency of our mental attitude makes the picture what it is, not the picture itself.

I want to end this chapter with a poem by Rudyard Kipling. This poem sums up the importance of the frequency of one's mental attitude. Read it. Then read it again. And again. You will begin to realize that no matter what "happens" to you, the quality of your life depends on how you respond to it - your attitude. If your outlook is one of enthusiastic high frequency, the circumstance is "good". If your outlook is one of pessimistic low frequency, the circumstance is "bad". *You* possess the power to look at the circumstances however you wish, and in the end, that is what will make all the difference.

[IF]
by Rudyard Kipling

If you can keep your head when all about you
are losing theirs and blaming it on you,
If you can trust yourself when all men doubt you
but make allowance for their doubting too,
If you can wait and not be tired by waiting,
Or being lied about, don't deal in lies,
Or being hated, don't give way to hating,
And yet don't look too good, nor talk too wise:

If you can dream--and not make dreams your master,
If you can think--and not make thoughts your aim;

If you can meet with Triumph and Disaster
And treat those two imposters just the same;
If you can bear to hear the truth you've spoken
Twisted by knaves to make a trap for fools,
Or watch the things you gave your life to broken,
And stoop and build 'em up with worn tools:

If you can make one heap of all your winnings
And risk it all on one turn of pitch-and-toss,
And lose, and start again at your beginnings
And never breathe a word about your loss;
If you can force your heart and nerve and sinew
To serve your turn long after they are gone,
And so hold on when there is nothing in you
Except the Will which says to them: "Hold on!"

If you can talk with crowds and keep your virtue,
Or walk with kings--nor lose the common touch,
If neither foes nor loving friends can hurt you;
If all men count with you, but none too much,
If you can fill the unforgiving minute
With sixty seconds' worth of distance run,
Yours is the Earth and everything that's in it,
And--which is more--you'll be a MAN my son!

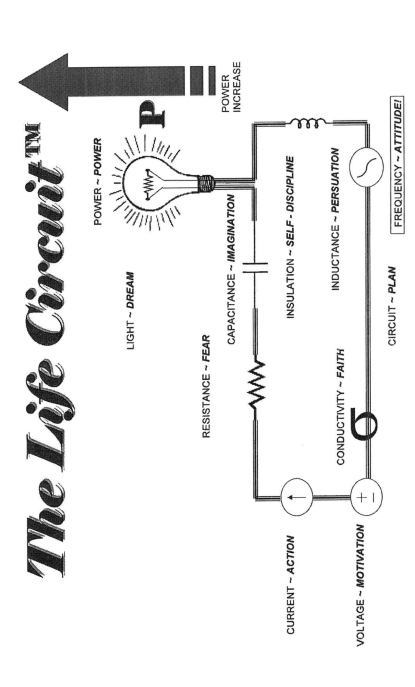

The Life Circuit™

POWER ~ **POWER**

POWER INCREASE

P

LIGHT ~ **DREAM**

CAPACITANCE ~ **IMAGINATION**

INSULATION ~ **SELF - DISCIPLINE**

INDUCTANCE ~ **PERSUATION**

RESISTANCE ~ **FEAR**

CONDUCTIVITY ~ **FAITH**

CIRCUIT ~ **PLAN**

CURRENT ~ **ACTION**

VOLTAGE ~ **MOTIVATION**

FREQUENCY ~ **ATTITUDE!**

CONCLUSION – COMPLETING THE CIRCUIT OF LIFE

The big question is, "If fear is what prevents us from taking action, how do we override that fear and act anyway?" Well, the answer is wrapped up in how badly you want to attain the very thing that your fear is blocking you from attaining. In other words, it all comes back to *your motivation.* This is the key.

As learned in Chapter Ten, an opening in the electric circuit will always prevent the electric current being pushed by the voltage source from crossing over to complete the circuit and turn on the light. The exception is when the voltage source pushing the electric current is an unnaturally powerful voltage source. In that case, the voltage source is so strong that it actually causes the electric current to *arc* across the opening in the electric circuit and flow to the other side of the circuit where it can complete the circuit.

I draw your attention now to Figure C1.

Figure C1 – Electrical Arc created by opening a 500kV disconnect switch

*I would like to thank Mr. Neil Brady for this photo. If Mr. Brady is reading this, please contact the author of this publication at: KolieCrutcher@ElectricLiving.net

Figure C1 shows a breath-taking electrical arc. The open circuit condition was created as a high-voltage disconnect switch opened at an electrical substation. Although there is an air gap created in the circuit because of the opening of the disconnect switch, the electric current is still being pushed by an incredible voltage. This voltage is so strong that it pushes the electric current through the opening in the circuit (for all practical purposes, air has an infinite resistance!), creating the magnificent electric arc across the open to complete the circuit. Not even the opening in the circuit can stop the electric current that is pushed by such a voltage source!

Ordinary motivation is not sufficient to attain the great successes of life. Our motivation (voltage source) must be so powerful, so unstoppable, that any open circuits (fears) in our life are no match for it. A voltage source of such magnitude simply pushes us so hard that we make the impossible possible, just as the electric current that arcs through the air to reach the other side of the circuit, turns on the light.

You must want success so badly, and with such intense desire that nothing stands in your way; with such an intense desire that when it appears that you are defeated, you are still pushed forward by an unseen source of motivation, and—like my man Monroe Mann—'romp on' regardless.

Achievement is measured by the magnitude of the fears we have conquered on the way to attaining our goal.

Former two-time world heavyweight boxing champion George Foreman once said that if he can clearly picture what he wants, he doesn't notice any pain getting to it. Celebrate your success! Even if you have not gotten there yet, celebrate your success by vividly envisioning your success. Etch a detailed picture of what you will do when you hit your goal, and focus on it. The clearer, brighter, and more colorful this picture, the less you will notice your fears and obstacles, and the sooner you will transmute your mental picture into reality!

There must be that undeniable force, that "something" that carries you when you feel you cannot carry yourself. It must be in tune with your dominant major purpose or dream in life. Life is more of a game than it is anything else, and just like any game there are rules that must be learned in order to play the game well. The difference between life and any other game, however, is that you do not have a choice as to whether you want to play the game of life. If you are on this earth, you are playing the Game of Life.

Many people do not understand this, and as a result try to sit on the sidelines of Life.

However, in Life, you are always still on the board. You must play! There are no other options.

Therefore, regardless of your situation, in order to achieve and succeed, you must complete the circuit and this includes "arcing across the open circuit" when the circuit of life decides to open on you, causing fear.

The only substantial difference between people lies in their ability to act in the face of their fears. Those who are great are not those without fear. Those who are great are those with fear but who have an unquenchable motivation backing them up that propels them forward *despite* that fear. Once an individual accepts the fact that he is afraid, but decides that there is something more important than that fear, he is on his way to Electric Living, and he will begin to create unstoppable power in his life.

As stated in the beginning of this book, no one can choose your motivation for you. As Monroe Mann says, "Romp On!" That is the attitude that signifies an intense motivation and desire for success, and is the battle cry of his band, *Running for Famous*. When you "Romp On", you never quit, you never stop, and not so much because you don't want to, but more because you just *can't*!

When you find that undeniable motivation—that motivation that has you possessed and obsessed with the attainment of your dream, you literally cannot stop. You press through all your fears and you use your obstacles to create the power necessary to turn on your Light. The power behind this attitude is the Power of Life.

That is Electric Living, and that is the Powerful Life!

Come to the edge, He said.
They said, we are afraid.
Come to the edge, He said.
They came.
He pushed them…and they flew.
- Guillaume Apollnaire

APPENDIX I

<u>BOOST YOUR FREQUENCY!!!</u>

GENIUS is the ability to assemble in new forms what already exists.
- Donald Trump

And the Lord answered me, and said, write the vision, and make it plain upon tables, that he may run that readeth it. For the vision is yet for an appointed time, but in the end it shall speak, and not lie: though it tarry, wait for it; because it will surely come, it will not tarry.
- Habakkuk 2:2-3

For most people, there is a gap between the person they perceive themselves to be, and the person they actually desire themselves to be. Once this gap is closed, true happiness occurs. The perceived self must "catch up" to the desired self through means of **<u>HUSTLE!</u>**
- Kolie E. Crutcher III

This "thing" that you are afraid of is not the issue. It does not exist! The issue is, can you act despite your fear.
- Kolie E. Crutcher III

And we know that all things work together for good to those that love God and are called according to His purpose for them.
- Romans 8:28

The biggest adventure you can ever take is to live the life of your dreams.
- Oprah Winfrey

The greatest skill a man can attain is the ability to act in the face of his fears.
- Kolie E. Crutcher III

100% of the shots you don't take don't go in.
- Wayne Gretzky

Double your failure rate!

- Thomas Watson

What happens to you is not important. How you react to it is.

- Pat Riley

There is nothing so disgusting as someone who has accepted his fate.

- Monroe Mann

Wealth is the progressive mastery of matter by mind.

- R. Buckminster Fuller

You Gotta Believe!

- Deion Sanders

Genius is 1% inspiration and 99% perspiration. Accordingly a genius is often merely a talented person who has done all of his or her homework.

- Thomas Edison

So God created man in His own image; in the image of God, He created him.

- Genesis 1:27

If it's hard, then do it hard.

- Les Brown

If we did all the things we are capable of doing, we would literally astound ourselves.

- Thomas A. Edison

You cannot discover oceans unless you have the courage to leave the shore.

- Successories

If you want to make enemies, try to change something.

- Woodrow Wilson

If you gave me 10 hours to accomplish a task, I would spend 9 hours planning and 1 hour executing it.

- Albert Einstein

Lack of opportunity is often nothing more than lack of purpose and direction.

- Unknown

The pessimist sees difficulty in every opportunity. The optimist sees opportunity in every difficulty.

- Winston Churchill

Whether you win or lose – it's how you play the Game.

- Kolie E. Crutcher III

I'M Possible

-Unknown

I can do all things through Christ who strengthens me.

- Philippians 4:13

The road to success is not straight. There is a curve called Failure, a loop called Confusion, speed bumps called Friends, red lights called Enemies, caution lights called Family. You will have flats called Jobs. But if you have a spare called Determination, an engine called Perseverance, insurance called Faith, and a driver named Jesus, you will make it to a place called Success.

- Author unknown

To know is nothing at all. To imagine is everything.

- Anatole France

Some of us seem to accept the fatalist position, the fatalist attitude, that the Creator accorded us to a certain position and condition, and therefore there is no need trying to be otherwise.

- Marcus Garvey

The wise man must be wise before, not after, the event.

- Epicharmus

Never, never, never give up!

- Winston Churchill

Most of the ailments we have – whether physical or mental – can be remedied by hard work.

- Kolie E. Crutcher III

Nothing great was ever achieved without enthusiasm.

-Ralph Waldo Emerson

The color of the rose lies within us, not in the rose.

-John Keats

Reality is wrong. Dreams are for real.

- Tupac Shakur

Success requires no explanations. Failure permits no alibis.

- Author unknown

A happy person is not a person in a certain set of circumstances, but rather a person with a certain set of attitudes.

- Hugh Downs

The best time to give it your all is when you've got nothing left to give.

- Monroe Mann

I never see failure as failure but only as the game I must play to win.

- Author Unknown

Why fit in, when you were born to stand out?

- Author Unknown

The successful person makes a habit of doing what the failing person doesn't like to do.

- Thomas Edison

Opportunity is missed by most people because it is dressed in overalls and looks like work.

- Thomas Edison

Think of yourself as on the threshold of unparalleled success. A whole, clear glorious life lies before you. Achieve! Achieve!

- Andrew Carnegie

I am not going to retire rich. I am going to be rich long before I retire.

- Monroe Mann

The only real limitation on your abilities is the level of your desires. If you want it badly enough, there are no limits on what you can achieve.

- Brian Tracy

I decided once and for all that I was going to make it or die.

- John H. Johnson

Achievement requires more than a vision – it takes courage, resolve and tenacity.

- Neil Eskelin

Ordinary people believe only in the possible. Extraordinary people visualize not what is possible or probable, but rather what is impossible. And by visualizing the impossible, they begin to see it as possible.

- Cherie Carter-Scott

You cannot climb a mountain if you will not risk a fall.

- Rick Beneteau

The greatest inventions and accomplishments began as the flicker of an idea. This tiny flame was then fueled by desire and faith. Watch out for those tiny little ideas. You have the potential to turn them into great things.

- Steve Brunkhorst

When a goal matters enough to a person, that person will find a way to accomplish what at first seemed impossible.

- Nido Qubein

You don't have to get it perfect – you just have to get it going. Babies don't walk the first time they try, but eventually they do.

- Mark Victor Hansen

The first step toward success is taken when you refuse to be a captive of the environment in which you first find yourself.

- Mark Caine

Courage is, with love, the greatest gift. We are, each of us, defeated many times – but if we accept defeat with cheerfulness, and learn from it, and try another way – then we will find fulfilment.

- Rosanne Ambrose-Brown

If one advances confidently in the direction of his dreams, and endeavors to live the life he has imagined, he will meet with a success unexpected in common hours.

- Henry David Thoreau

The secret of success in life is for a man to be ready for his opportunity when it comes.

- Benjamin Disraeli

Something in human nature causes us to start slacking off at our moment of greatest accomplishment. As you become successful, you will need a great deal of self-discipline not to lose your sense of balance, humility, and commitment.

- H. Ross Perot

There is no scarcity of opportunity to make a living at what you love; there's only a scarcity of resolve to make it happen.

- Dr. Wayne Dyer

Success is waking up in the morning, whoever you are, wherever you are, however old or young, and bounding out of bed because there's something out there you love to do, that you believe in, that you're good at — something that's bigger than you are, and you can hardly wait to get at it again today.

- Whit Hobbs

There comes that mysterious meeting in life when someone acknowledges who we are and what we can be, igniting the circuits of our highest potential.

- Rusty Berkus

When you get to the end of all the light you know, and it's time to step into the darkness of the unknown, faith is knowing that one of two things shall happen: either you will be given something solid to stand on, or you will be taught how to fly.

- Edward Teller

Nothing will work unless you do.

- Maya Angelou

Cherish your visions and your dreams. They are the children of your soul, the blueprints of your ultimate achievements.

- Napoleon Hill

Optimism is the faith that leads to achievement. Nothing can be done without hope or confidence.

- Helen Keller

I believe life is constantly testing us for our level of commitment, and life's greatest rewards are reserved for those who demonstrate a never-ending commitment to act until they achieve. This level of resolve can move mountains, but it must be constant and consistent. As simplistic as this may sound, it is still the common denominator separating those who live their dreams from those who live in regret.

- Tony Robbins

APPENDIX II

RECOMMENDED READING LIST

(Leaders are Readers)

- *The Bible*
- *Think and Grow Rich* by Napoleon Hill
- *Think and Grow Rich: A Black Choice* by Dennis Kimbro
- *Do You!* by Russell Simmons w/ Chris Morrow
- *Guaranteed Success* by Percy Miller (Master P)
- *The Secret* by Rhonda Byrne
- *Never Mind Success – Go For Greatness! The Best Advice I Ever Received* by Tavis Smiley
- *The Theatrical Juggernaut* by Monroe Mann
- *Battle Cries for the Underdog* by Monroe Mann
- *Guerrilla Networking* by Jay Conrad Levinson & Monroe Mann
- *The Artist's MBA* by Monroe Mann
- *Time Power* by Brian Tracy
- *Focus* by Al Ries
- *The Magic of Thinking Big* by Dr. David Schwartz
- *Rich Dad, Poor Dad* by Robert Kiyosaki
- *Cash Flow Quadrant* by Robert Kiyosaki
- *Building Wealth One House at a Time* by John W. Shaub
- *Words of Wisdom: Daily Affirmations of Faith* by Rev Run
- *Becoming Rich* by Mark Tier
- *Rhinoceros Success* by Scott Alexander
- *Guerrilla Marketing* by Jay Conrad Levinson
- *Guerrilla P.R.* by Michael Levine
- *Titan (John D. Rockefeller)* by Ron Chernow
- *The Wizard of Menlo Park* by Randall Stross
- *The 48 Laws of Power* by Robert Greene
- *Blink: The Power of Thinking Without Thinking* by Malcolm Gladwell
- *How to Advertise* by Kenneth Roman and Jane Maas
- *Lewis Latimer: Creating Bright Ideas* by Eleanor H. Ayer

- *How to Think Like Einstein* by Scott Thorpe
- *Electricity Demystified* by Stan Gibilisco
- *Teach Yourself Electricity and Electronic (4th Edition)* by Stan Gibilisco
- *The Complete Idiot's Guide to Physics (2nd Edition)* by Johnny T. Dennis and Gary Moring
- *Electricity – A Self-Teaching Guide* by Ralph Morrison
- *Electrical Engineering 101 – Everything You Should Have Learned in School...But Probably Didn't* by Darren Ashby
- *Basic Electricity & DC Circuits* by Charles W. Dale, Ed.D
- *Schaums's Easy Outlines Applied Physics* by Arthur Beiser, Ph.D
- *Schaum's Outline of Basic Electricity (2nd Edition)* by Milton Gussow, M.S.
- *Basic Electricity (Revised Edition)* by Van Valkenburgh, Nooger & Neville, Inc.

TIISG (Turn It Into Something Good) Chart

"BAD" CIRCUMSTANCE	WHY "BAD"	HOW TO TIISG

All reader comments are sincerely welcomed and greatly appreciated. It is you—the reader—that brings value to the words within these pages. Without you, these words are nothing more than black ink on white pages. Thank you!

To contact the author regarding consulting services and speaking engagements, or if you just have something you would like to share, please feel free to contact me—Kolie E. Crutcher III—via email at:

KolieCrutcher@ElectricLiving.net

and visit my website:

www.ElectricLiving.net

So until we meet, I leave you with this great truth of life so eloquently spoken by Nelson Mandela at his 1994 Inauguration:

Our deepest fear is not that we are inadequate. Our deepest fear is that we are powerful beyond measure. It is our light, not our darkness that most frightens us. We ask ourselves, who am I to be brilliant, gorgeous, talented and fabulous. Actually, who are you not to be?

You are a child of God. Your playing small doesn't serve the world. There's nothing enlightened about shrinking so that other people won't feel insecure around you.

We were born to make manifest the glory of God that is within us. It's not just in some of us; it's in everyone. And as we let our own light shine, we unconsciously give others permission to do the same thing.

As we are liberated from our own fear, our presence automatically liberate others.